Interference Mitigation in Device-to-Device Communications

Interference Mitigation in Device-to-Device Communications

Edited by

Masood Ur Rehman
James Watt School of Engineering, University of Glasgow, UK

Ghazanfar Ali Safdar
School of Computer Science and Technology, University of Bedfordshire, UK

Mohammad Asad Rehman Chaudhry
Soptimizer, Toronto, Canada

WILEY

This edition first published 2022

Registered Offices
John Wiley & Sons, Inc., 111 River Street, Hoboken, NJ 07030, USA

John Wiley & Sons Ltd, The Atrium, Southern Gate, Chichester, West Sussex, PO19 8SQ, UK

Editorial Office
The Atrium, Southern Gate, Chichester, West Sussex, PO19 8SQ, UK

For details of our global editorial offices, customer services, and more information about Wiley products visit us at www.wiley.com.

Wiley also publishes its books in a variety of electronic formats and by print-on-demand. Some content that appears in standard print versions of this book may not be available in other formats.

Library of Congress Cataloging-in-Publication Data

Names: Ur Rehman, Masood, editor. | Safdar, Ghazanfar Ali, editor. | Chaudhry, Mohammad Asad Rehman, editor.
Title: Interference mitigation in device-to-device communications / edited by Masood Ur Rehman, Ghazanfar Ali Safdar, Mohammad Asad Rehman Chaudhry.
Description: Hoboken, NJ : Wiley, 2022. | Includes index.
Identifiers: LCCN 2021046319 (print) | LCCN 2021046320 (ebook) | ISBN 9781119788799 (cloth) | ISBN 9781119788805 (adobe pdf) | ISBN 9781119788812 (epub)
Subjects: LCSH: Wireless communication systems. | Electromagnetic interference.
Classification: LCC TK5103.2 .I5179 2022 (print) | LCC TK5103.2 (ebook) | DDC 621.384–dc23
LC record available at https://lccn.loc.gov/2021046319
LC ebook record available at https://lccn.loc.gov/2021046320

Cover Design: Wiley
Cover Image: © metamorworks/Shutterstock

Set in 9.5/12.5pt STIXTwoText by Straive, Chennai, India

10 9 8 7 6 5 4 3 2 1

Dedicated to ALLAH, the Lord and the Sustainer of all the realms.
Asad lowers his wings of humility to Mum and Abbu.
Mohammad Asad Rehman Chaudhry
Masood Ur Rehman
Ghazanfar Ali Safdar

To my biggest inspiration, my parents, Khalil Ur Rehman and Ilfaz Begum
my support system, siblings; Habib, Waheed, Tahera and wife; Faiza
and apple of my eyes, my son, Musaab.
Masood Ur Rehman

To my late parents, Safdar Hussain and Sadiq Sultana
and my two little angels, Taha and Taqi and my wife, Misbah.
Ghazanfar Ali Safdar

Contents

Preface

Emergence of data-extensive applications such as online gaming and video sharing has resulted in exponential increase in the mobile data traffic. Advances in Internet-of-Things (IoT) and 5G require fusion of multiple sensors and wireless devices operating in real time. Supporting such high data rate demands within the framework of existing wireless access networks is a challenging task. Addressing this ever-increasing demand of data hungry devices in an efficient and effective manner has driven the wireless industry to investigate new paradigms. Device-to-device (D2D) communication is a promising solution to this complex problem and hence a key-enabling technology for 5G. The D2D is envisioned to operate either in out-band mode representing usage of a dedicated spectrum or in-band mode demonstrating operation within the same spectrum of the existing cellular spectrum.

The D2D communication provides the closely located users with an opportunity to communicate directly without traversing traffic through the base station while offering exciting advantages of improved throughput, increased spectrum reuse, and enhanced energy gain. The attractive features and simple infrastructure of D2D communication are making it experience a tremendous growth, having an expectation of the global number of D2D connections specifically to reach 14.7 billion devices by 2023. Applications of D2D communication cover a wide variety of fields ranging from communication to infotainment, healthcare to surveillance, and public safety to manufacturing.

Integration of the D2D communication with legacy systems brings new technical challenges. Interference to the primary users is one of the major challenges that needs to be managed and mitigated effectively in order to have an optimal system performance. The current developments and expected future growth of the D2D communication necessitate a comprehensive reference dealing with this fundamental and paramount issue.

A dedicated book that addresses these important issues is rarely available. This book is, therefore, a laborious effort to put together the basis of interference and principles for advanced management techniques in D2D communication. It brings together multidisciplinary contributors in the field of wireless communication, system modeling, and signal processing to take up on the challenges and devise solutions related to the development, integration, and enhancement of D2D systems to solicit an interference-free operation.

The state of the art as well as advanced topics on interference mitigation in standard as well as emergency scenarios are discussed. Interference management through power control, resource allocation, spectrum splitting, MIMO antennas, interference manipulation, user pairing, content delivery, subchannel assignment, reinforcement learning, mode selection, and clustering is considered. Moreover, a comprehensive account of system modeling, design challenges, and performance analysis are presented. Particular focus is put on the promising new dimension of D2D-based disaster management and use of Unmanned Aerial Vehicles for infrastructure-less communications. The book is concluded by highlighting potential future application areas and avenues of research and development for D2D communication. Treatment of variety of topics makes this book a valued reference for the communication system designers, researchers, lecturers, and students.

Masood Ur Rehman
Ghazanfar Ali Safdar
Mohammad Asad Rehman Chaudhry

Acknowledgments

Editing a book is more strenuous than we thought and more rewarding than we could have imagined. This journey took two years for us to enjoy this sense of contentment. We would like to thank the commissioning editor Sandra Grayson, managing editor Juliet Booker, editorial assistant Becky Cowan, and publishers who showed great patience and extended a supporting hand throughout the journey of compiling this book. None of this would have been possible without their help.

We would also like to thank Mr. Khuram Ashfaq and Mr. Abduljabbar for their help in reviewing Chapters 1 and 10, respectively.

And finally, our sincere gratitude goes to all the contributors whose participation resulted in an excellent reference on this interesting topic.

About the Editors

Masood Ur Rehman received the BSc degree in Electronics and Telecommunication Engineering from University of Engineering and Technology, Lahore, Pakistan, in 2004, and the MSc and PhD degrees in Electronic Engineering from Queen Mary University of London, London, UK, in 2006 and 2010, respectively. He worked at Queen Mary University of London as a Postdoctoral Research Assistant till 2012 before joining the Centre for Wireless Research at University of Bedfordshire as a Lecturer. He served briefly at the University of Essex and then moved to the James Watt School of Engineering at University of Glasgow in the capacity of an Assistant Professor in 2019. His research interests include compact antenna design, radiowave propagation and channel characterization, satellite navigation system antennas in cluttered environment, electromagnetic wave interaction with human body, body-centric wireless networks and sensors, remote healthcare technology, mmWave and nano communications for body-centric networks, and D2D/H2H communication. He has worked on a number of projects supported by industrial partners and research councils. He has contributed to a patent and authored/co-authored 4 books, 8 book chapters, and more than 120 technical articles in leading journals and peer-reviewed conferences.

He is a Fellow of the Higher Education Academy (UK), a Senior Member of the IEEE, a Member of the IET, and a part of the technical program committees and organizing committees of several international conferences, workshops, and special sessions. He is acting as an Associate Editor of the IEEE Access, IET Electronics Letters & Microwave and Optical Technology Letters and lead guest editor of numerous special issues of renowned journals. He also serves as a reviewer for book publishers, IEEE conferences, and leading journals.

Ghazanfar Ali Safdar received his BSc Hons in Electrical Engineering from University of Engineering and Technology, Pakistan, and MEngg in Computer Science and Telecommunications from ENSIMAG, INPG, France. He was awarded a PhD from Queen's University Belfast, UK, in 2005, for his work in the area of power-saving MAC protocols for the IEEE 802.11 family of wireless LANs. Moreover, he has worked as a Research Fellow at Queen's University Belfast on a project related to Wireless Networks security. Currently, he is working as a Senior Lecturer in Computer Networking at the University of Bedfordshire, UK.

His research interests mainly include cognitive radio networks, energy-saving MAC protocols, security protocols for wireless networks, LTE networks, interference mitigation, device-to-device communications, network modeling, and performance analysis. He has authored/co-authored 4 books, 13 books chapters, and around 100 plus research articles in leading journals and peer-reviewed conferences.

He is an Associate Fellow of Higher Education academy (UK) and a Session Chair and part of technical program committees of several international conferences. He has also worked as R&D Engineer with Carrier Telephone Industries (SIEMENS), Pakistan, and Schlumberger, France. He is currently serving as the Editor-in-Chief of EAI Endorsed Transactions on Energy Web and Information Technology, and an Area Editor for Springer Wireless Networks. He regularly serves as a reviewer for several esteemed journals, book proposals, and conferences.

Mohammad Asad Rehman Chaudhry received a PhD in Electrical and Computer Engineering from Texas A&M University, and an MBA from University of Toronto-Rotman School of Management. He is a thought-leader, strategist, innovator, and entrepreneur. He has been leading multidisciplinary global projects in Digital Disruption and its adaptation for Socio-Economic transformation. He has delivered several keynotes at international events. He has advised senior executives from Fortune 500 companies on Digital Strategy and Future-Tech. He leads IEEE standards related to Software Defined and Virtualized Ecosystems. He is also a co-founder of two companies. He has previously held positions with IBM Research, Hamilton Institute, University of Toronto, DARPA System F6, and the University of Calgary. He has served in various executive and board positions. He is a recipient of the Fulbright Fellowship and the Presidential Scholarship.

List of Contributors

Sohail Ahmed
Department of Avionics Engineering
Air University
Islamadad
Pakistan

Kamran Ali
Computer Science Department
Middlesex University
London
United Kingdom

Rafay I. Ansari
Department of Computer and Information Sciences
Northumbria University
Newcastle
United Kingdom

Zakia Asad
Soptimizer
Canada

Muhammad Awais
Computer Science Department
Edge Hill University
Ormskirk
United Kingdom

Mohammad Asad Rehman Chaudhry
Soptimizer
Canada

Imran Haider
Acme Center for Research in Wireless Communications (ARWiC)
Capital University of Science and Technology
Islamabad
Pakistan

Muhammad A. Imran
James Watt School of Engineering
University of Glasgow
Glasgow
United Kingdom

Byungtae Jang
Electronics and Telecommunications Research Institute
Daejeon
South Korea

Tahera Kalsoom
School of Computing, Engineering and Physical Sciences
University of West of Scotland
Paisley
United Kingdom

Nauman Javed
School of Electrical Engineering & Computer Science
National University of Sciences and Technology
Islamabad
Pakistan

Rida Khan
Thomas Johann Seebeck Department of Electronics
Tallinn University of Technology
Tallinn
Estonia

Aboubaker Lasebae
Computer Science Department
Middlesex University
London
United Kingdom

Hassan Malik
Department of Computer Science
Edge Hill University
Ormskirk
United Kingdom

Muhammad Y. M. Mirza
Lahore Business School
University of Lahore
Lahore
Pakistan

Mujahid Muhammad
Birmingham City University
Birmingham
United Kingdom

Bushrah Naeem
Department of Computer Science
Balochistan University of Information Technology
Engineering, and Management Sciences (BUITEMS)
Quetta
Pakistan

Huan Nguyen
Computer Science Department
Middlesex University
London
United Kingdom

Tien H. Nguyen
School of Electronics and Telecommunications
Hanoi University of Science and Technology (HUST)
Hanoi
Vietnam

Van D. Nguyen
Hanoi University of Science and Technology
Hanoi
Vietnam

Xuan T. Nguyen
School of Electronics and Telecommunications
Hanoi University of Science and Technology
Hanoi
Vietnam

Vishnu V. Paranthaman
RDS Global Ltd.
Derby
United Kingdom

Haris Pervaiz
School of Computing and Communications
Lancaster University
Bailrigg
United Kingdom

Farrukh Pervez
School of Electrical Engineering & Computer Science
National University of Sciences and Technology
Islamadad
Pakistan

Asad A. Pirzada
School of Electrical Engineering & Computer Science
National University of Sciences and Technology
Islamadad
Pakistan

Mohsin Raza
Computer Science Department
Edge Hill University
Ormskirk
United Kingdom

Ghazanfar Ali Safdar
School of Computer Science and Technology
University of Bedfordshire
Luton
United Kingdom

Purav Shah
Computer Science Department
Middlesex University
London
United Kingdom

Navuday Sharma
Test Software Development
Ericsson Eesti AS
Tallinn
Estonia

Muhammad N. Sial
Department of Informatics
King's College London
London
United Kingdom

Anum Tanveer
Computer Science Department
Middlesex University
London
United Kingdom

Masood Ur Rehman
James Watt School of Engineering
University of Glasgow
Glasgow
United Kingdom

Yanli Xu
College of Information Engineering
Shanghai Maritime University
Shanghai
China

Daeseung Yoo
Electronics and Telecommunications Research Institute
Daejeon
South Korea

Taehyun Yoon
Electronics and Telecommunications Research Institute
Daejeon
South Korea

Kamran Zia
School of Electrical Engineering & Computer Science
National University of Sciences and Technology
Islamabad
Pakistan

1

Introduction to D2D Communications

Ghazanfar Ali Safdar[1], Masood Ur Rehman[2], and Mohammad Asad Rehman Chaudhry[3]

[1] *School of Computer Science and Technology, University of Bedfordshire, Luton, United Kingdom*
[2] *James Watt School of Engineering, University of Glasgow, Glasgow, United Kingdom*
[3] *Soptimizer, Canada*

A perceptible upsurge has been discerned in wireless data traffic in recent years due to the occurrence of numerous new applications, thereby imposing an abrupt requirement for a large capacity network. Among them, multimedia applications such as high definition (HD) video streaming, HD video conferencing, and online gaming are most common [1]. Another factor impacting wireless data traffic is the primer of smart wearable technology such as watches, air pods and wrist band. In the near future, these devices would increase tenfold or even more, and their presence ensures more and more usage of smart wireless devices. For majority of the users, the wireless cellular network is the main access system for the internet due to its universal availability. According to Cisco white paper [2], there will be 12.3 billion mobile-connected devices by 2022 and mobile will represent 20 percent of total IP traffic, and global mobile data traffic will increase sevenfold between 2017 and 2022, reaching 77.5 exabytes per month by 2022. This pretense new contest to those working in the field of cellular communication and operators is deploying more and more wireless links between terminals with radio networking to meet the required demand.

1.1 D2D Communication

Existing device-to-device (D2D) technologies such as Bluetooth and Wi-Fi direct are available for thorough communication between two devices in close contiguity. They can associate different types of devices since the level of compatibility between modern devices is very high. The pairing procedure is considerably easy and both technologies simplify the entire pairing process by making enabled devices readily discoverable to one another [3]. Devices can communicate over a

Interference Mitigation in Device-to-Device Communications, First Edition.
Edited by Masood Ur Rehman, Ghazanfar Ali Safdar, and Mohammad Asad Rehman Chaudhry.

reliable communication channel with a moderate date rate and latency. Since the communication range is limited in these technologies, mobile device benefits from low power consumption that improves the battery life of the devices. In addition, it allows different devices to connect and communicate with each other without the need of an additional infrastructure.

Despite all the above benefits, both technologies have some limitations like single hop communication, limited range, latency, and moderate data. Both technologies are operated in the unlicensed band; hence, interference is eminent due to the number of devices already operated in that band [4]. Also, on an unlicensed band, the maximum transmit power constraint confines communication to a limited range. Both technologies have asynchronous nature, and the device receiver is supposed to continuously monitor the channel. The reason for this channel monitoring (idle listening) is not to miss the discovery signal from other devices, but this can drain the device battery significantly. Finally, to establish a reliable communication, the users are responsible for the discovery and initiation mechanism, and there is no centralized control. All of the above factors are the main obstacles in the integration of both technologies with next generation of mobile networks and they fail to act as a D2D on large scale. Another most common technology operated in the unlicensed band is Wi-Fi that has already assisted cellular networks in offloading local traffic and easing congestion. But its implementation requires a setup of access points connected to the internet and performance degrades as the number of user increases.

The diversity of radio accessing in mobile users provides lots of flexibility for D2D communication in terms of link establishment, resource allocation, energy efficiency, as well as applications and services. D2D communication enables researchers to merge together the achievements of long-term development in previously two disjoint networking techniques, i.e. ad-hoc networking and centralized networking. D2D users can form clusters or multihop routes, operates autonomously or under partial or full control of the operator. On the other hand, with the assistance of the operator, it is possible to achieve network-wide performance optimization after properly allocating frequency bands to D2D pairs and managing the interference there. D2D communication in cellular band is a very flexible communication technique with unique advantages over existing D2D techniques. D2D in cellular band can significantly enhance the cellular communication in terms of system capacity, coverage, throughput, latency, and user experience [1]. D2D communication can offload traffic from the serving base station by quickly exchanging a large amount of data among mobile devices in short range and alleviate congestion in the cellular core network.

In addition to that, D2D can also support local area services such as content distribution, local advertisement, and location aware services very effectively through unicast, groupcast, and broadcast transmission. A D2D in-coverage user can also act as a relay and extend the existing coverage of a serving base station and benefits the device that is on the edge of the cell boundary to experience better coverage [1]. The serving base station has the overall view of the entire cell, and it can facilitate D2D communication in different aspect, for example, finding the neighboring devices, reducing the transmitting power, and providing a suitable environment for communication. D2D implementation in cellular network does not require additional infrastructure and hence suppresses the high implementation costs. Undoubtedly, the reduction in transmission delay and offloading traffic from the existing base station to eases congestion in the cellular core network are the driving factors toward its addition in next generation of mobile networks [4].

In recent research, there are various attractive benefits of D2D in terms of spectral efficiency, energy efficiency, through put, and signal-to-interference and noise ratio (SINR). However, it is still unknown that if these gains could be fully utilized in the practice. In D2D communication, in spite of having many advantages, there are still many open challenges for the successful implementation of this technology. In particular, D2D communication will require efficient device discovery mechanisms (among devices in close proximity), intelligent mode selection (D2D or cellular mode) algorithms, complex resource management techniques (to avoid interference), and robust security protocols. Interference management is one of the compelling areas in D2D communication, and researchers have proposed different schemes, protocols, and methods to reduce interface. It is quite beneficial from a capacity and spectral efficiency point of view that D2D users reuse the same cellular resource for D2D communication. But it generates interference and requires different techniques to ensure that the quality of service (QoS) of a cellular link is not degraded because of D2D communication.

1.2 Evolution of D2D Communication

The wireless revolution in telecommunication begun in 1990s with the introduction of cellular network and has been driven by advances in radio frequency (RF) and microwave engineering [11]. In the past, infrared (IR) wireless technology was used for point-to-point links that use devices that are inexpensive, compact, lightweight, and consume low power. The limitations of IR technology were short distance, mobility of devices during transmission, and line-of-sight (LOS) requirement between transmitter and receiver. Unlike IR, the RF technology was effectively used for a long-distance communication that can penetrate through obstacles and does not require a LOS. RF technology operates in the range of 3 kHz–300 GHz and used for communication and broadcasting. The radio spectrum is divided into different segments and assigned to different technology industries, for example, very high frequency (VHF) band that ranges from 30 to 300 MHz is assigned to FM radio and TV broadcast, while ultra-high frequency (UHF) is dedicated to cellular, wireless local area network (WLAN), and Bluetooth, etc.

Current D2D technologies such as Bluetooth and Wi-Fi direct operate in unlicensed industrial, scientific, and medical (ISM) band; therefore, the interference with other devices operated in similar band is uncontrollable. Both suffers from limitations such as single hop communication, interference, limited range, latency, moderate data rate, and no centralized control. The burden of device discovery and communication initiation is also upon the user to establish a successful and

Table 1.1 Comparison between D2D technologies.

Technology/ parameters	**Device-to-device in LTE-A**	**Bluetooth**	**Wi-Fi direct**
Standardization	3GPP LTE-A (release 12)	Bluetooth SIG 5.0	IEEE 802.11, WECA
Frequency band	Licensed band for LTE-A	2.4 GHz	2.4 and 5 GHz
Max transmission distance	Up to 500 m	Up to 200 m	Up to 200 m
Max transmit power	24 dBm	4 dBm	15 dBm
Data throughput	1 Gbps	1.4 Mbps	250 Mbps
Frequency sharing	In-band UL or DL frequency spectrum	Frequency hoping spread spectrum (FHSS)	Orthogonal frequency division multiplexing (OFDM)
Modulation technique	SC-frequency division multiple access (FDMA) (UL), orthogonal frequency division multiple access (OFDMA) (DL)	GFSK	DSSS
Latency	D2D mode – Ultra-low Cellular mode – 9 ms	200 ms	150 ms
Power consumption	Low in D2D mode High in cellular mode	Low	High
Interference	Controlled interference environment that leads to less interference	Yes	Yes
Quality of service	Provide hard QoS guarantees	Bluetooth medium access control (MAC) scheduling	Wi-Fi multimedia extensions (WME)
Security	Better security guarantee	Service-level security Device-level security	Strong Wi-Fi protected access (WPA)
Control	Centrally control by the base station (either fully or partially)	No centralized control	No centralized control
Discovery procedure	Discovery via broadcast	Use SDP (service discovery protocol) to locate services that are available nearby	Two steps asynchronous message-based discovery

Table 1.1 (Continued)

Technology/ parameters	Device-to-device in LTE-A	Bluetooth	Wi-Fi direct
Hardware requirement	ProSe application sever incorporated in evolved packet core (EPC) ProSe application installed on user device	Bluetooth adaptor on all the devices connecting with each other	Wireless adaptors on all the devices
Implementation complexity	High – ProSe application on User device ProSe application server incorporated in EPC	Low – Small low-cost microchip embedded in Bluetooth enabled device	Low – It is relatively simple and requires configuration of hardware and software.
Applications	Offload traffic, multimedia content sharing, local advertising and PS applications	Data sharing between peripheral connection	Content sharing, group gaming and device connection
Primary devices	Cellular device only	Mobile phones, mouse, keyboards, office, and industrial automation devices	Cellular devices, portable media players, printers, cameras, and scanners

reliable communication. The biggest obstacle is the inherit architecture of both technologies that does not allow a centralized controller to control and manage the D2D communication. Future cellular nodes would have the global view of the entire cell, and they can effectively manage the available resources for a successful D2D communication. The diversity of cellular communication provides lots of flexibility for D2D communication in terms of link establishment, resource allocation, energy efficiency, as well as applications and services. The communication on a licensed cellular band can be better in terms of interference avoidance under a controlled environment.

D2D communication can be utilized during the absence of an active mobile network, and it allows the device to establish an alternate wireless link with the surrounding in-coverage devices. This can increase the overall coverage area and D2D user acting as a relay extends communication over a greater range. Other applications of D2D are content distribution, local advertisement, and location-aware services [5]. D2D can also be implemented in machine-to-machine communication where one machine can talk with another machine in close proximity. D2D can also be implemented in vehicle-to-vehicle communication and other autonomous driving applications. A comprehensive comparison between Bluetooth, Wi-Fi direct, and D2D in cellular band is shown in Table 1.1

1.3 D2D Communication in Cellular Spectrum

D2D communication has always been present in the unlicensed spectrum but it was not investigated in the licensed spectrum in the first three cellular generations. The idea of D2D in licensed spectrum was introduced in the fourth generation and third generation partnership project (3GPP) announced D2D functionality in release 12 in 2012. In order to embed D2D functionality in long term evolution-advanced (LTE-A) network, they proposed a proximity services (ProSe) function to exchange data and voice among nearby devices. Initially, the discovery procedure was limited to in-coverage scenario where a user device autonomously selects the resource from a preconfigured pool. In addition, D2D communication was also proposed for public safety (PS includes police, firefighter, and ambulances) services in an area with no active network coverage [6, 7]. Later in release 13 in 2014, a further enhancement in D2D was proposed that include one-to-one communication between devices and discovery procedure was extended to out-of-coverage scenario. ProSe function was also proposed to use as a user equipment (UE)-to-network relay by means of a layer 3 based relaying and UE acts as a relay between a remote UE and a base station.

Since release 14, the implementation of D2D application is focused on V2X (that includes Vehicular to Vehicular, Vehicular to Infrastructure, Vehicular to Network, and Vehicular to Pedestrian) and aims to guarantee a safety application for comfort driving with high degree of quality and reliability (Table 1.2). In release 15, the D2D communication is extended to platooning (group of vehicles that operates like a train with virtual string attached between vehicles), extended sensors (that perform functions like exchanging live videos among nearby vehicles, road site units, pedestrian devices, and V2X application services), advance driving (enables semi or full automated driving), and remote driving.

In release 16, the focus is on the two iterations of the LTE-V2X, 3GPP system is adding advanced features primarily in the area of low latency use cases.

Table 1.2 Use cases for D2D in 3GPP release.

Release 12	Release 13	Release 14	Release 15	Release 16 and beyond
PS	PS enhancement	V2V	Platooning	Maritime communication system
ProSe	ProSe enhancement	V2I	Extended sensors	Side link enhancements include V2X
	UE-to-network relay	V2N	Advanced driving	Critical communications
		V2P	Remote driving	Commercial use between smartphone directly

It introduces a concept of "Side-link enhancements" for the D2D communication between devices, this is primarily used for the automotive use case as well as for critical communications. There is also a push toward using side-link for commercial use between smartphone directly. The 3GPP system in release 16 also focus on the development of a maritime communication system known as LTE-Maritime. The overall coverage area will be up to 100 km, and there will be seamless communication between the existing 3GPP system and the maritime radio communication system. It will also support voice communication and data communication between different vessels at sea [8, 9]. This intervehicular communication can occur using D2D communication technology.

V2X communication capabilities are undergoing rapid development to enhance the experience of comfort and safe self-driving. 5G cellular communication system is evolving to provide support to V2X because of its higher data rate, low latency, reliability, global deployment, and broader coverage. 3GPP has specified V2X services in LTE-A network in release 14 and 15 and introduced a concept of enhanced V2X in release 16. Vehicles not only can communicate with each other through multiple wireless hops but also through V2I links. The V2I link can improve the communication stability and reduce the reaction time of the drivers. In order to ensure timely vehicular safety communications, higher data rate is required to achieve fast data exchanges that can significantly improve the interworking of V2V and V2I. V2X can also help in the detection of traffic congestion because of real time information, and it does not require any additional infrastructure or roadside equipment. However, V2X implementation is not as simple as it looks it can encounter different real time traffic scenarios that are caused by varying driving speeds, traffic patterns, and driving environment and poses greater challenges (Figure 1.1).

D2D communication in cellular spectrum enables direct communication between devices that are in close proximity, and it is an exciting and innovative feature for next generation of cellular networks (Figure 1.1). It can increase spectral and energy efficiency, provides high data rate, low latency, controlled interference, as well as offloads traffic from the base station and alleviate congestion [6, 10] (Figure 1.2). It also provides higher control because of centralized management and allows D2D communication in a disciplined environment. D2D functionality was not implemented in the first four generations of mobile communication as this was only considered as a tool to reduce the cost of local services.

Moreover, D2D can play a critical role in areas where infrastructure is damaged by natural disasters, i.e. earthquake or hurricane. An urgent communication network can easily be set up in a very short time scale that can benefit PS services. Previously in emergency situation, ad hoc networks were created with the help of local devices that communicate with each other. Despite several benefits of ad hoc networks, it lacks connectivity to the existing communication network [4]. D2D users in emergency situations can act as relay and supports communication over a greater range hence increasing the overall coverage area of cellular network. D2D technology can also be very instrumental in disaster hit areas for issuing alerts and updates using local content distribution feature.

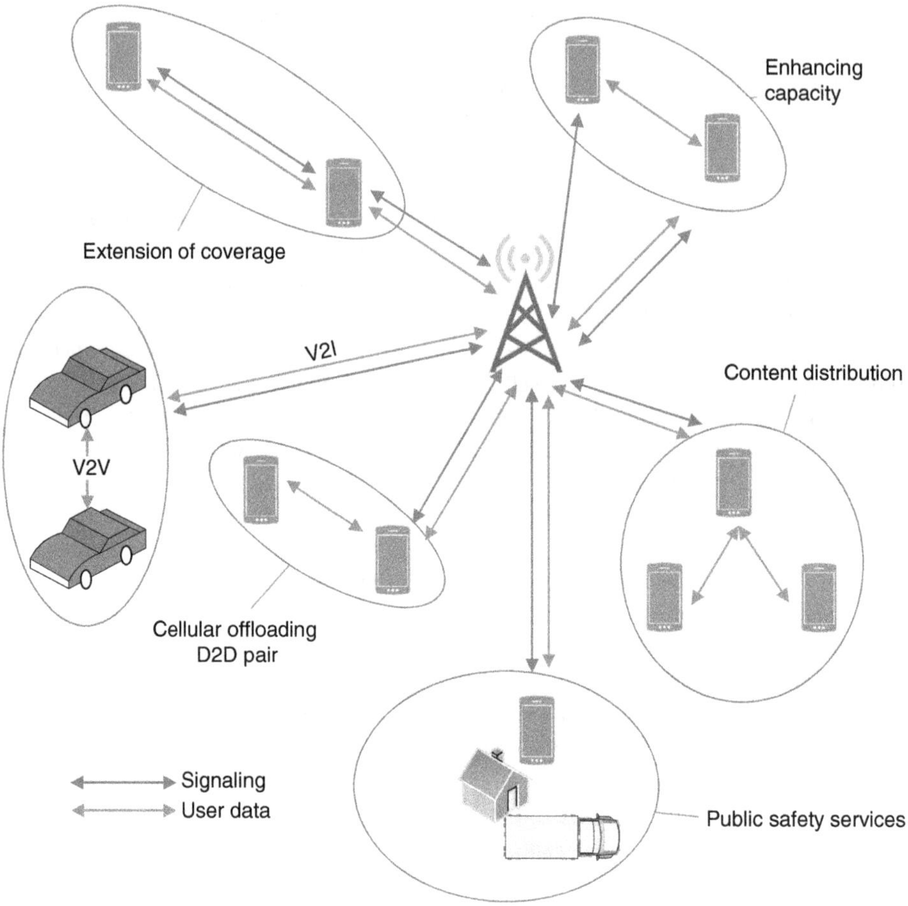

Figure 1.1 Device-to-device communication modes.

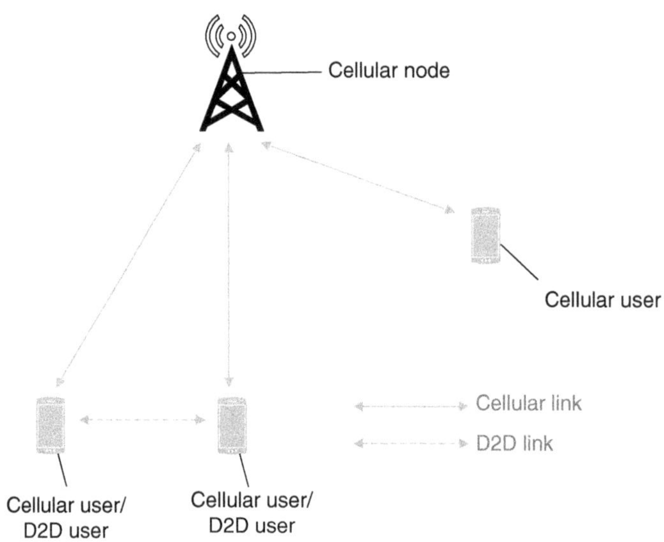

Figure 1.2 D2D communication overview.

1.4 Classification of D2D Communication

In terms of spectrum usage, D2D is primarily classified into two categories, in-band and out-band [6]. In in-band, D2D user shares the licensed spectrum band with the cellular user and managed by the serving base station. In-band D2D, can be an underlay (re-use model) where eNodeB shares radio resource between D2D and cellular user or overlay (fixed resource) where eNB allocates a fixed resource for D2D. In out-band, D2D user uses the unlicensed ISM frequency band for D2D and can either be controlled by eNB (network controlled) or by the D2D user them self (autonomous) [6, 10]. The main advantage of out-band D2D is that it eliminates the interference problem linked to cellular user when using licensed band.

A. In-band D2D Communication
   In in-band D2D communication, D2D user shares the cellular licensed spectrum used in either uplink (UL) or downlink (DL) communication [12]. This approach increases the spectral efficiency of the existing cellular network and is further divided into underlay (non-orthogonal) and overlay (orthogonal) modes (Figure 1.3).
B. Out-band D2D Communication
   In out-band, D2D communication uses unlicensed ISM band where cellular communication does not occur (Figure 1.4). The biggest advantage is the elimination of interference between D2D and cellular users, although we still face interference from other devices that are operating in that band, for example, Bluetooth and Wi-Fi direct users. Out-band D2D is further categorized into controlled and autonomous mode, in controlled mode, the radio interface for D2D is controlled by the cellular network, while in autonomous mode, the D2D user control the link for D2D. In this scheme, resource allocation is much simpler as eNodeB is not considering exiting cellular RB, time, and location. Moreover, additional infrastructure enables a user to maintain both cellular and D2D connection at the same time, but this results in poor power consumption

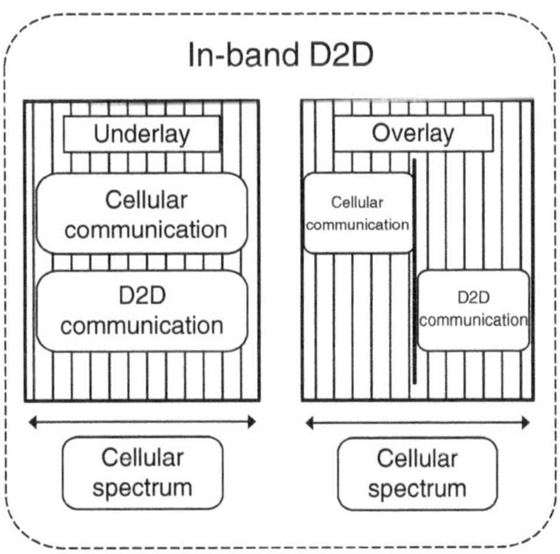

Figure 1.3 In-band D2D communication.

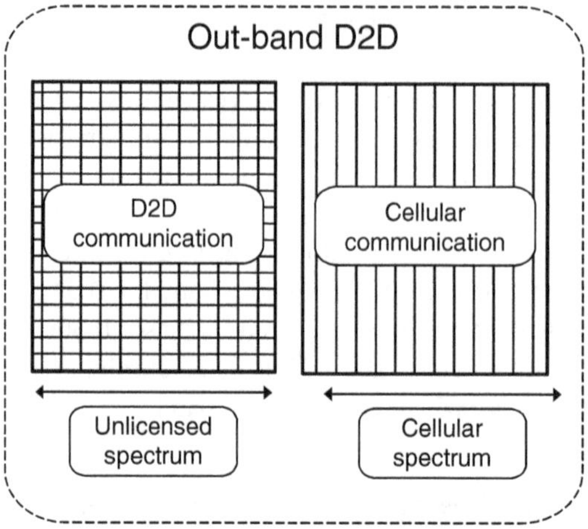

Figure 1.4 Out-band D2D communication.

and leads to uncontrollable intersystem interference in unlicensed band. The D2D in unlicensed band can also increase congestion and poor QoS for exiting Wi-Fi direct and Bluetooth users and can affect overall network performance. This scheme is not widely backed by the researchers because of security issues, dual radio interface, and power management. Hence, in-band D2D is given much attention and is most likely the way ahead in terms of D2D functionality and its implementation in next generation of mobiles communication.

1.5 Challenges in D2D Implementation

Local area services are considered as essential part of future networks, but unlicensed spectrum reuse led to inefficient utilization of resources and a compromise in the QoS because of decentralised controller [13]. Therefore, embedding D2D in exiting LTE-A technology is the most appropriate scheme for effective D2D communication. D2D in in-band underlay in LTE-A network provides fast access to the spectrum band with controlled interference by the network infrastructure [6, 9, 10].

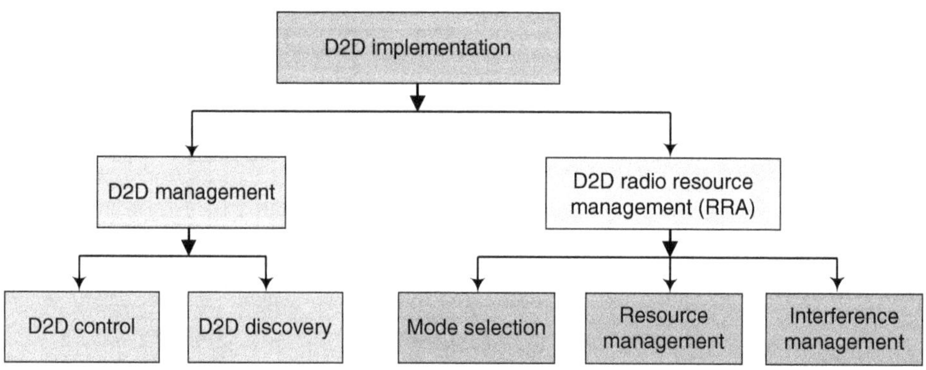

Figure 1.5 Challenges in D2D implementation.

It not only can provide higher spectral efficiency but also achieve different types of gains such as proximity, hop, reuse, and pairing. Although, D2D provides benefit in terms of performance improvement, spectral, and energy efficiency, but there are still few challenges while integrating D2D in LTE-A network. These challenges are broadly categorized into D2D management (D2D control and device discovery) and D2D radio resource management (RRM) (mode selection, resource management, and Interference), Figure 1.5.

1.6 Summary

D2D communication is a likely technology that targets at capitalizing on system throughput through improved spectrum efficiency. Nevertheless, spectrum re-claim results in detrimental interference among the primary cellular users and the secondary D2D users. This unwanted interference damages the inclusive network performance, which must be undertaken. This requires the need to further investigate this area, in order to proficiently alleviate cross-tier and co-tier interference during resource sharing. Accordingly, the rest of the book chapters 2–9 are aligned to address the core issues around D2D technology.

References

1 Jameel, F., Hamid, Z., Jabeen, F. et al. (2018). A survey of device-to-device communications: research issues and challenges. *IEEE Communication Surveys and Tutorials* 20 (3): 2133–2168.

2 Cisco Annual Internet Report (2018–2023) White Paper, 2020, Cisco. https://www.cisco.com/c/en/us/solutions/collateral/service-provider/visual-networking-index-vni/white-paper-c11-738429.html (accessed 21 October 2021).

3 Miller, M. (2001). *Discovering Bluetooth*. Sybex Inc.

4 Sauter, M. (2017). *From GSM to LTE-Advanced Pro and 5G*. Wiley.

5 Shahid, M. and Jonathan, R. (2014). *Smart Device to Smart Device Communication*. Springer.

6 Technical Specification Group Radio Access Network (2013). Evolved Universal Terrestrial Radio Access (E-UTRA) and Evolved Universal Terrestrial Radio Access Network (E-UTRAN); Overall Description; Stage 2 (Release 11), 3GPP TS 36.300 V11.7.0, September 2013.

7 Fodor, G., Dahlman, E., Mildh, G. et al. (2012). Design aspects of network assisted device-to-device communications. *IEEE Communications Magazine* 50 (3): 170–177.

8 Zhang, B., Li, Y., Jin, D. et al. (2015). Social-aware peer discovery for D2D communications underlaying cellular networks. *IEEE Transactions on Wireless Communications* 14 (5): 2426–2439.

9 C. F. M. e Silva, T. F. Maciel, R. L. Batista et al. (2015). Network-assisted neighbor discovery based on power vectors for D2D communications. *2015 IEEE 81st*

Vehicular Technology Conference (VTC Spring), pp. 1–5. https://doi.org/10.1109/VTCSpring.2015.7145949.

10 Huan, T., Ding, Z., and Levy, B. (2014). Enabling D2D communications through neighbor discovery in LTE cellular networks. *IEEE Transactions on Signal Processing* 62 (19): 5157–5170.

11 Ur-Rehman, M. and Safdar, G.A. (editors) (2018). LTE communications and networks: Femtocells and antenna considerations. Wiley & Sons (UK).

12 Safdar, G.A., Ur-Rehman, M., Muhammed, M. et al. (2016). Interference mitigation in D2D communication underlaying LTE-A network. *IEEE Access* 4: 7967–7987.

13 Ashfaq, K., Safdar, G.A., and Ur-Rehman, M. (2021). Comparative analysis of scheduling algorithms for radio resource allocation in future communication networks. *PeerJ Computer Science* 7: e546.

2

Interference Mitigation in D2D Communication Underlaying LTE-A Network

Ghazanfar Ali Safdar[1], *Masood Ur Rehman*[2], *Mujahid Muhammad*[3], *and Muhammad A. Imran*[2]

[1] *School of Computer Science and Technology, University of Bedfordshire, Luton, United Kingdom*
[2] *James Watt School of Engineering, University of Glasgow, Glasgow, United Kingdom*
[3] *School of Computing and Digital Technology, Birmingham City University, Birmingham, United Kingdom*

The ever-increasing popularity of smart devices (smartphone, tablet PC, etc.) and portable applications designed with various mobile applications that provides multimedia-rich services, e.g. Video over IP, online game, Internet Protocol Television (IPTV), etc., have directly increased the demand for high mobile data traffic. This results in the wireless cellular network to be one of the main access systems to the Internet, for majority of customers, due to its pervasive availability. Thus, today's mobile data traffic consumes a lot of bandwidth within the cellular network. For instance, it was estimated that global mobile data traffic would experience 13-fold rise from 2012 to 2018 [1, 2].

Since the evolution of mobile communication standards 2G, 3.5G, 4G and now in the 5G era, the wireless cellular network has been integrating advanced technologies to improve data rates, network capacity, and coverage. The 3GPP groups have proposed LTE-A standard to be an enhanced version of LTE, which aims at supporting peak data rate in the order of Gbits/s. To achieve this performance, the LTE-A have been specified to integrate emerging technologies such as multi-antenna technique and carrier aggregation to provide higher channel capacity, device-to-device (D2D) communication, etc. D2D communication is a radio access technology that provides users the ability to communicate directly between them when they are in close proximity, without traversing traffic through the network infrastructure [3]. It is envisaged as an import part of 5G and beyond networks and Internet-of-Things (IoT) applications [4, 5]. Traditionally, in conventional cellular network, all data transmission traverses the base station also referred to as evolved Node B (eNB), i.e. data packets are first uploaded to the eNB using the uplink resources and then routed to the intended destination along the downlink [6]. However, this traffic generates a significant overhead to the eNB, as it provides services to many mobile users. With the implementation of D2D, data transmission can follow a direct communication channel, established between two devices that are in close proximity. This, therefore, offload data traffic from the eNB, providing it with ability

Interference Mitigation in Device-to-Device Communications, First Edition.
Edited by Masood Ur Rehman, Ghazanfar Ali Safdar, and Mohammad Asad Rehman Chaudhry.

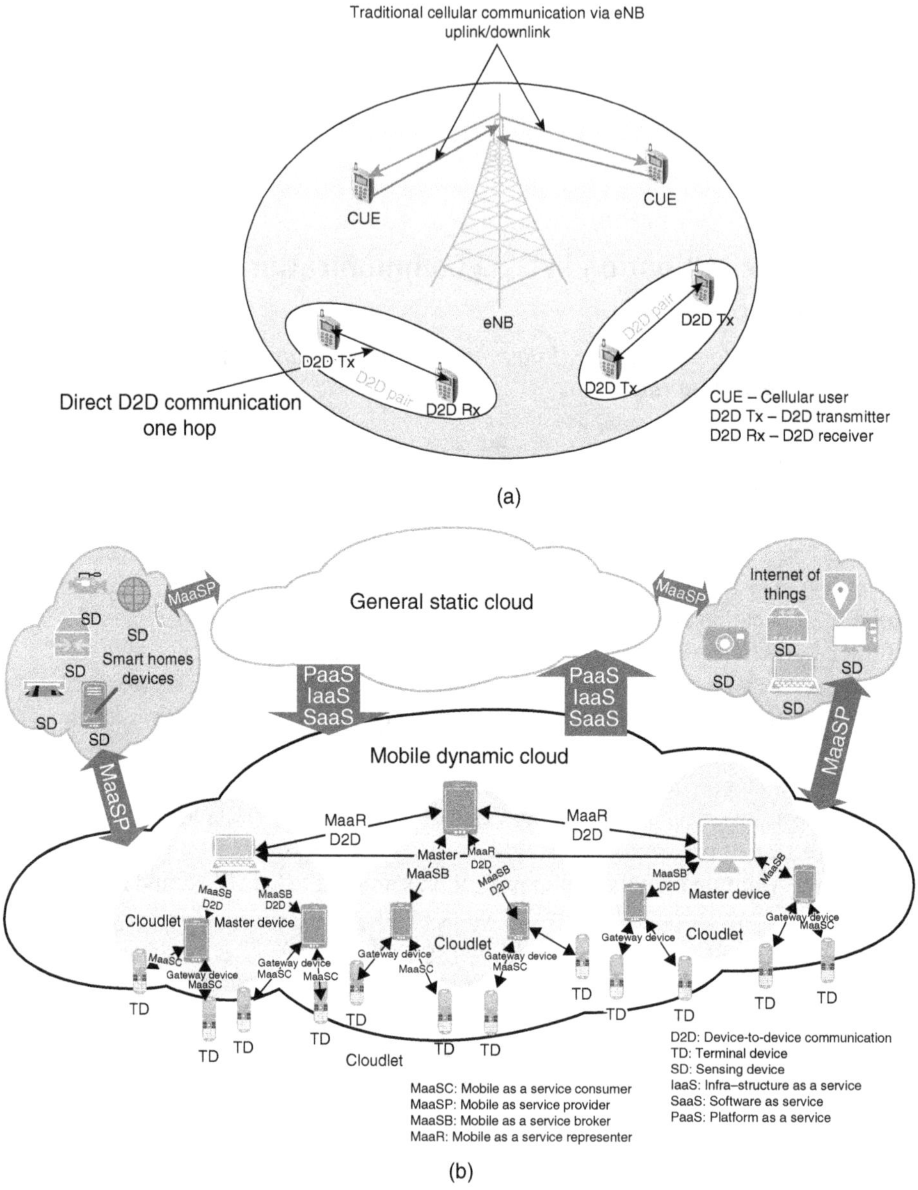

Figure 2.1 D2D communication architecture [7]. (a) D2D communication as an underlay to LTE-A network. (b) D2D-based heterogeneous network architecture for mobile cloud computing.

to serve other devices which are not in close proximity and are using the same channel, thus easing network congestion and increase in network capacity. For two devices to operate in D2D mode, they must satisfy proximity conditions. Proximity here implies not only close range but also includes good channel condition above a predefined signal-to-interference and noise ratio (SINR) target, average throughput, delay, density, and load [6].

Figure 2.1 illustrates the D2D communication architecture. Direct communication between D2D pair, i.e. D2D transmitter (D2D Tx) and D2D receiver (D2D Rx) nodes that are in close proximity, using the direct link between them, without passing traffic through the eNB, as in traditional cellular transmission can be seen in Figure 2.1a. Although both D2D Tx and D2D Rx are transceiver devices that operate in half-duplex mode, but for simplicity, the device that first initiate data transmission is referred to as D2D Tx, while the receiving device is termed as D2D Rx. Therefore, we follow the same naming convention in this chapter.

2.1 Applicability of D2D Communication

The applicability of D2D spans across many areas including, but not limited to proximity-based services, e.g. social application, smart communication between vehicles, content distribution, multicasting, peer-to-peer communication, location-aware advertisement, and public safety (disaster management and rescue operations). D2D is currently being specified in 3GPP LTE-Release 12 and is recognized as one of the technologies components of future wireless networks (i.e. 5G) [8]. Figure 2.2 shows the various use cases and applicability of D2D-enabled cellular system.

D2D communication in cellular network promises several benefits to the network providers as well as end users. Firstly, it improves spectral efficiency, by reusing the same frequency resources occupied by cellular users. Secondly, it reduces communication delay and increases network throughput, due to its short-distance

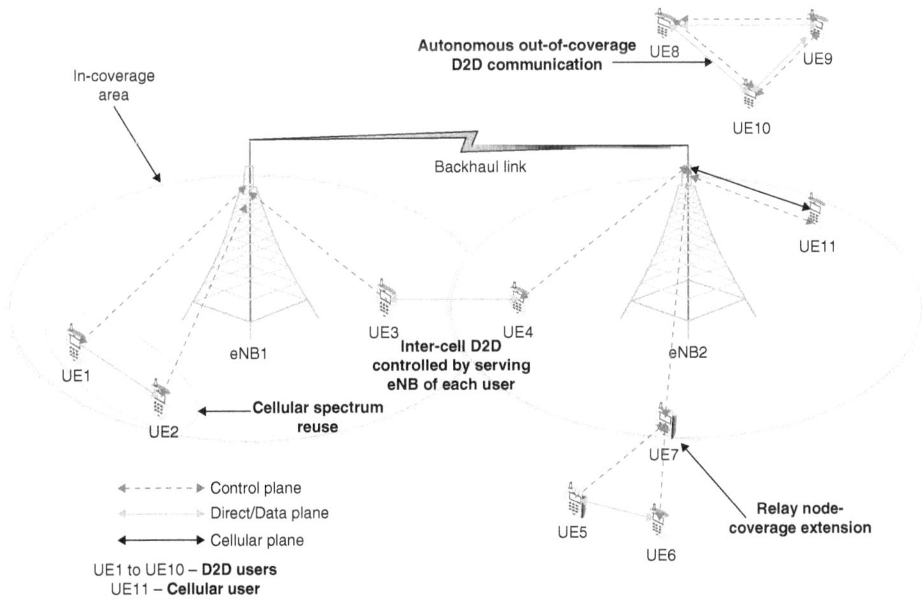

Figure 2.2 Illustration of D2D-enabled cellular network concepts and use cases [7].

communication. Thirdly, D2D improves energy efficiency since the mobile terminals use less transmission power when communicating directly between each other. Hop gain is another key benefit of using D2D, in which either the uplink or downlink resources is only used for the direct communication link.

From economic perspectives, the advent of D2D drives the network application developers to design new mobile applications with proximity-aware features such as social content, multicasting applications, etc. [9–11]. The mobile operators will also benefit from an increased revenue generation, through additional charges for providing D2D services to subscribers, using the same licensed cellular spectrum band. However, this is still an open-research question, especially when considering how the operators will control and charge for D2D communications (same operator and different operators scenarios), and also how can the relay devices be compensated using their resources (battery life, data storage, system processing power, etc.) [12]. Therefore, an efficient and reliable pricing model needs to be developed and implemented for the benefits of both operators and user devices that my act as relay terminals for D2D communication. Furthermore, organizations can benefit from location-based advertisement, through adverting their product and services as part of discovery information to nearby devices that are in close proximity. The end user would benefit from an improved data rates, reduced latency due to direct short-range communication, and overall improved user experience.

D2D communication has been introduced in the heterogeneous mobile network architecture to enhance the efficiency of mobile cloud computing services as well as the conventional cloud computing. The benefit of such adoption is to decentralize the computationally intensive big cloud into distributed small mobile clouds (cloudlets), which can be managed by mobile devices with high processing capability. The significance of D2D communication in mobile cloud computing and a new hierarchical heterogeneous architecture consisting of D2D relay-based cloud services was described in Ref. [13]. As shown in Figure 2.1b, the cloudlets offload data traffic from the general static cloud provide services to its associated cloud members (e.g. sensing devices, terminal devices). Various cloudlets are interconnected through a direct D2D link from their individual master devices. These therefore utilize the benefits of D2D communication to improve load balancing, energy efficiency, user experience, and reduced computational overhead. On the other end, the existing mobile cloud computing models i.e. are used to provide services between the general static cloud and the mobile dynamic cloud.

On the other hand, enabling D2D communication in existing cellular network opens up new technical challenges that need to be tackled. These are device discovery, mode selection, interference management, radio resource management (RRM), power control, security, mobility management, and modification to the existing architectural design of LTE-A network. The aforementioned challenges have motivated a lot of research studies toward, proposing optimal solutions to achieve efficient and reliable D2D-enabled cellular system. Detailed analysis of these challenges is presented in the subsequent sections.

2.2 Interference – The Compelling Issue in D2D

Although the integration of D2D communication in cellular network would offer enormous benefits such as higher achievable data rates and spectrum efficiency, improved capacity, ease of congestion, reduced power consumption, etc; however, it may subject the primary cellular users to harmful interference due to spectrum sharing, in addition to mutual interference among the multiple D2D pairs reusing the time–frequency resources. The presence of undesirable interfering signals could result in performance degradation of the communications in wireless networks generally [14] and in cellular and the D2D communication system, specifically. In other words, enabling D2D in cellular system should not cause service disruption to cellular users (CUs). Therefore, efficient interference coordination schemes need to be implemented to guarantee the quality of service (QoS) constraint of both transmission systems, and to also provide reliable communication environment. In this regards, various interference avoidance techniques have been studied in the literature such as transmit power control [15–17], efficient RRM schemes [15, 16], joint power control and resource allocation [17, 18], multiple-input multiple-output (MIMO) techniques [19, 20].

The conventional power control strategy alone cannot handle the interference problem in this heterogeneous network where D2D pairs coexist with cellular network, because there is the presence of new interference situation due to simultaneous spectrum sharing. Whereas mitigating the interference problem in D2D-enabled cellular network through the optimization of diverse resource-sharing algorithms limits the benefits of D2D reuse. This results in low throughput due to inefficient use of bandwidth resources. Consequently, the application of MIMO transmission scheme at cellular downlink to mitigate the interference caused by D2D communication does not exploit the full potentials of D2D communication in cellular network. This is because; it does not cater for interference from cellular users to the D2D Rx in the UL period, when CU and D2D pair shares the same UL resources. This necessitates the need to further investigate this area in order to efficiently eliminate the interference problems in D2D communication.

Therefore, to protect the cellular network from D2D interference and improve the reliability of both D2D communication and cellular networks, there is need to suppress both D2D interference and cellular interference during the UL period, when both D2D pairs and cellular users share the same UL resources. This can be achieved by exploiting beamforming strategy at the D2D pair, in which the precoding vector can be aligned toward the direct link between the D2D pair, and at the same time nulling interference to the cellular users.

2.3 Types of D2D Communication

Basically, there are two types of D2D communication, namely, in-band and out-band. The major difference between the two is the frequency spectrum band in

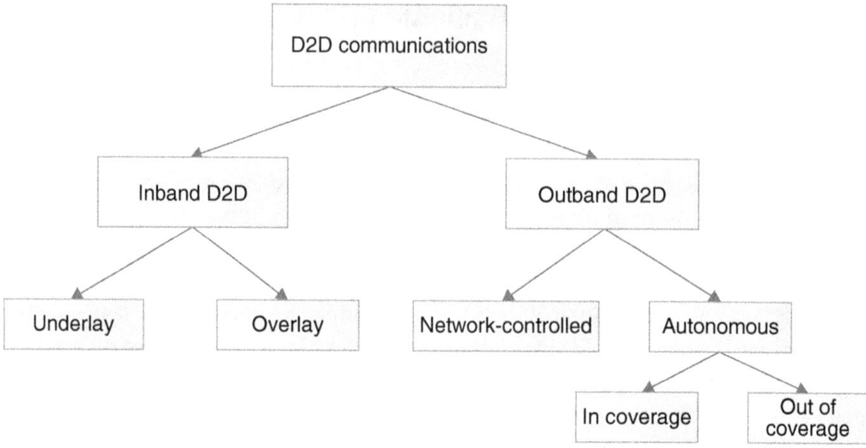

Figure 2.3 Classification of D2D communication [7].

which the D2D communication is operating. Figure 2.3 shows a block diagram illustrating the available classes and subclasses of D2D communication.

2.3.1 In-Band D2D Communication

In in-band communication, D2D share licensed cellular spectrum along with other cellular users in the LTE-A network. The network infrastructure, i.e. eNB, has total/partial control over the D2D users. The eNB is responsible for discovering potential D2D devices, link establishment based on channel state information (CSI), radio resource allocation either uplink or downlink, power control based on some certain predefined threshold level as well as interference coordination between the cellular and D2D users [21]. The in-band D2D communication is further divided into underlay (non-orthogonal) and overlay (orthogonal) modes.

2.3.1.1 Underlay In-Band

In underlay in-band, the D2D users and cellular users are allocated the same radio resources simultaneously by the eNB. The eNB reuses either the uplink or downlink resource blocks for D2D communication based on certain performance metrics such as mutual distance between D2D and cellular users, transmit power level, interference-limited area, etc. This type of implementation is also known as the reuse mode (non-orthogonal resource sharing). The reuse mode can achieve higher spectrum efficiency compared with overlay. However, it introduces severe interference problem between the D2D and cellular users, since both users are simultaneously using the same physical resource blocks [22, 23]. Since the cellular users are the primary users of the spectrum band, underlaying D2D communication on the same band implies that QoS of the cellular communication as well as the D2D itself becomes a challenge [16]. Furthermore, the reliability issue of both communications needs to be addressed. Several literatures exist in this area, focusing on providing novel approaches in solving these aforementioned challenges.

2.3.1.2 Overlay In-Band

In this type of D2D communication also referred to as dedicated mode, fixed (orthogonal) resources are allocated for D2D communication by the eNB from the cellular band. For this reason, the mutual interference between cellular and D2D users is eliminated, since each communication mode occupy a separate and dedicated physical resource blocks for their individual communication [24]. But still, there is mutual interference among the D2D users, since multiple D2D links can reuse the same RBs for their transmissions, which affect the overall network throughput. The overlay in-band has a major drawback of underutilization of radio resources as compared to underlay mode. This is because when there is no any D2D session, the dedicated resources are left ideal, which translate to inefficient spectrum usage [6]. Therefore, underlay D2D is gaining more popularity in the industry as well as the academia due to its higher spectral efficiency than overlay. Figure 2.4 illustrates random distribution of D2D and cellular users in a single-cell scenario, where some D2D pairs reuse CU's uplink (UL) resources, while others use non-overlapping resources, for D2D communication.

2.3.2 Out-Band D2D Communication

Out-band D2D communication exploits the unlicensed ISM frequency band for its operations [25]. This is similar to the operating band of Wireless Local Area Network (WLAN) and Bluetooth technologies. In out-band D2D, coordination and management of D2D connections can either be controlled by the eNB also known

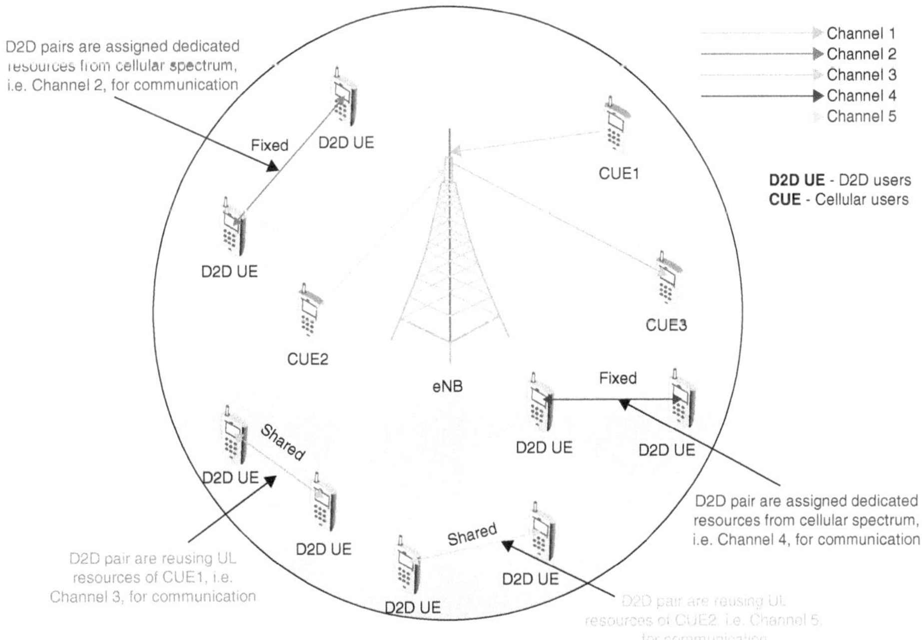

Figure 2.4 A single cell comprising of multiple cellular links, and D2D links using different D2D communication modes [7].

as *network-controlled*, or by the D2D users themselves, referred to as *autonomous*. The main advantage of this category of D2D is that it eliminates the interference problem between cellular and D2D links. Also, resource allocation becomes easier since the scheduler (i.e. eNB) does not need to take into account frequency, time, and location of users when assigning resource blocks (RBs) to both D2D and cellular users.

Furthermore, users can simultaneously maintain cellular and D2D connections using the two radio interfaces. However, the major drawback is its uncontrollable inter-system interference due to the presence of other communicating entities, e.g. Wi-Fi and Bluetooth devices that operate in the same unlicensed band. Thus, sharing unlicensed spectrum might not provide a stable controllable environment and would lead to congestion and poor QoS experience and affect the overall network throughput. In addition, security of D2D transmission [13] and coordinating communication over two different bands with independent radio interfaces incurs crucial power management problem [26]. For these reasons and other related issues, researchers and mobile operators pay much attention to in-band D2D communication, specifically underlay. Therefore, this chapter will focus on underlay in-band D2D communication. Figure 2.5 shows schematic frequency band occupancy of D2D communication in cellular networks.

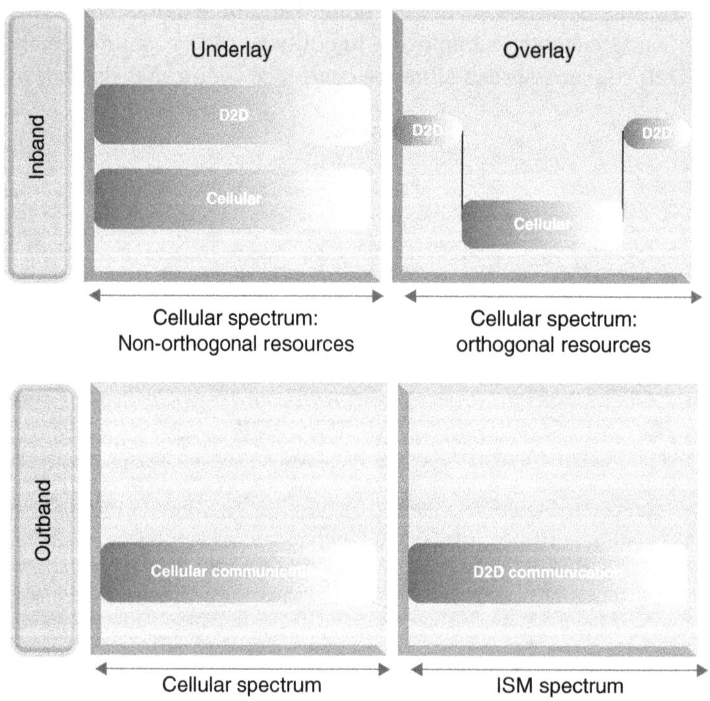

Figure 2.5 Schematic frequency band occupancy of D2D communication in cellular networks [7].

Out-band D2D communication can be categorized either as network-assisted (network-controlled) or autonomous D2D depending on the level of network involvement in managing and coordinating D2D communication.

2.3.2.1 Network-Assisted D2D Communication

In network-assisted *D2D communication*, the eNB is responsible for synchronizing D2D users in time, frequency, and phase using primary synchronization signal (PSS) and secondary synchronization signal (SSS) during cell search procedure. It provides control information signals via physical downlink control channel (PDCCH), for device discovery, session setup, link establishment, resource scheduling assignment, power control, routing, etc. Also, the eNB monitors the D2D links to ensure that the D2D policies are not violated. The D2D users periodically feedback status report on direct link and other control information surrounding the environment, to the eNB via physical uplink shared channel (PUSCH), random access channel (RACH), etc. Such report information includes CSI, SINR, device discovery request, scheduling request, and so on [27]. In essence, the eNB has total control of all D2D link activities and performs RRM effectively, and hence, any undesirable interfering signals both from the cellular and D2D communications can easily be coordinated. Network-assisted D2D provides the benefits of satisfying the QoS requirements of cellular communication while managing D2D communications effectively and efficiently, with the aim of improving the overall system throughput. However,

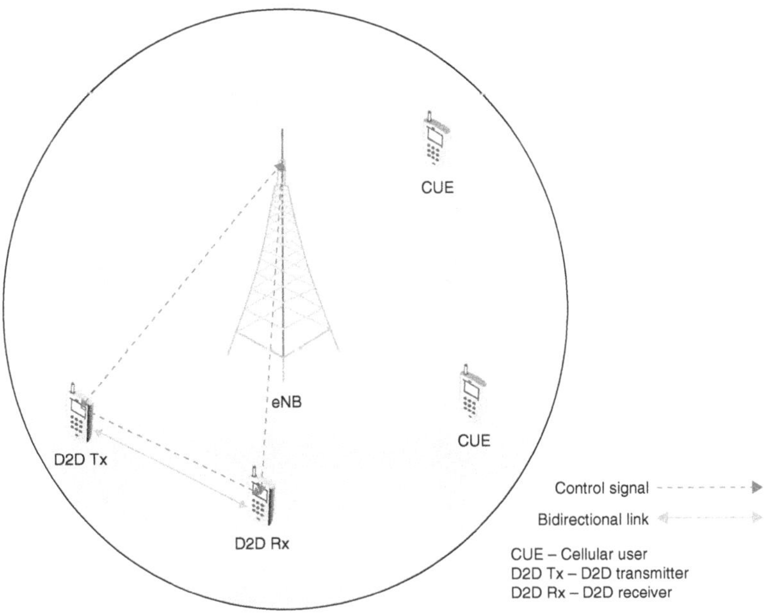

Figure 2.6 Network-assisted D2D control message exchange and link establishment procedure [7].

it incurs high signaling overhead necessary to manage and control D2D activities. For example, the eNB requires the knowledge of the full CSI of all involved links for interference avoidance techniques. This increases complexity on the part of the eNB. Figure 2.6 demonstrates a network-assisted D2D for a single-cell scenario, where the D2D procedures are managed by the eNB.

2.3.2.2 Autonomous D2D Communication

In autonomous D2D communication, the eNB has partial control over the activities of the D2D users or links. The eNB performs radio resource allocation in a large timescale and put restriction on the maximum transmit power allowed at the side of D2D users, etc. D2D users independently establish communication session through direct discovery, by announcing and monitoring process between the D2D pair, as shown in Figure 2.7. At the same time, the D2D users are able to control radio resource allocation and schedule their own transmission and set power control autonomously in a distributed fashion. Autonomous D2D communication is applicable in both in-coverage and out-of-coverage area scenarios where there is serving eNB and no cellular network infrastructure, respectively. In both cases, the D2D pair can independently establish communication between each other. The key benefit of autonomous D2D is that the eNB incurs less signaling overhead, which allows it to serve other cellular users. However, interference management among D2D users and high implementation complexity on part of the D2D users are among the major challenges for the autonomous D2D. Table 2.1 summarizes the key differences between the in-band and out-band D2D communication.

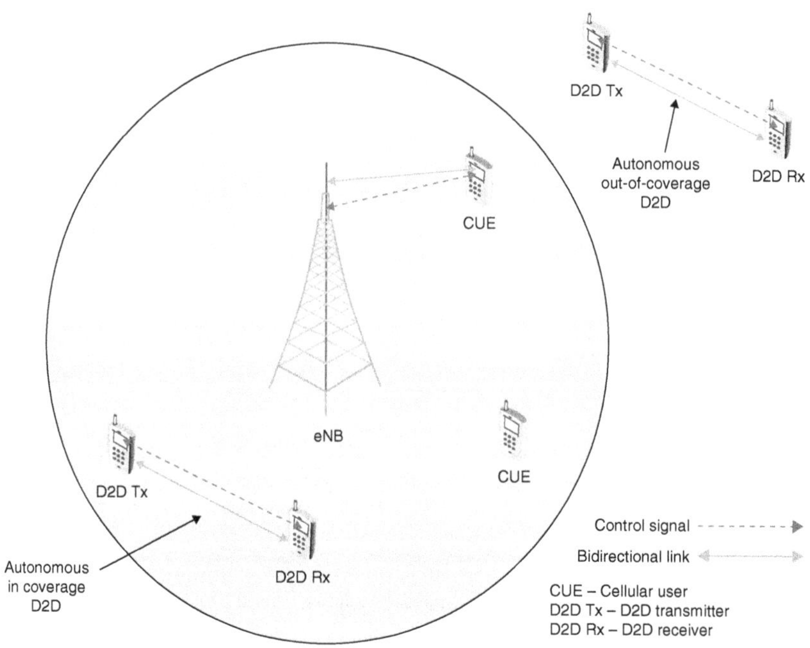

Figure 2.7 Device-controlled discovery and session setup procedures [7].

Table 2.1 Comparative analysis of D2D communication.

	In-band		Out-band	
Criteria	**Overlay**	**Underlay**	**Network-assisted**	**Autonomous**
Spectral efficiency	Medium	Very high	Low	Low
Interference between cellular and D2D users	Very low	High	Negligible	Negligible
Interference among D2D users	Medium	Medium	High	High
Controlled interference environment	Yes	Yes	No	No
Implementation complexity	Medium	Low	High	High
Extra signaling overhead to network	Medium	High	High	Very low
Inter-platform coordination	No	No	Yes	Yes
Network-controlled	Full/hybrid	Full/hybrid	Full	Loose
Cell coverage	In coverage/ partial coverage	In coverage/ partial coverage	In coverage	In coverage and out of coverage
Simultaneous D2D and cellular transmission	No	No	Yes	Yes
Energy efficiency	High	High	Low	Low

Source: Safdar et al. [7].

2.4 D2D Communication Underlaying Cellular Network – The Challenges

D2D communication as an underlay in LTE-A network enables fast access to the spectrum band with controlled interference by the network infrastructure. With underlay implementation, D2D communications can provide higher spectral efficiency and network throughput, which are the two main requirements for the LTE-A network. Also, D2D aims to achieve four types of gains, namely proximity, hop, reuse, and pairing. Proximity gain is achieved through short direct range communication via D2D links, which results in high data rate, low latency, and low power consumption. Second is the hop gain, where D2D transmission uses only one hop, as compared to traditional cellular communication via eNB, in which both uplink and downlink resources are used. Reuse gain is achieved by simultaneously allocating the same radio resources to both D2D and cellular links. Last is the pairing gain, which enables new type of wireless local area services, e.g. social services, and a UE can choose between cellular and D2D communication modes. Song et al. indicate that with D2D communication, the overall network throughput may rise up to 65% compared with the situation in which all the D2D traffic follow conventional cellular mode [19]. In addition, the D2D operation, i.e. pairing, handover, etc., can be fully transparent to the users, as compared to WLAN or Bluetooth technologies. Although D2D communication provides several benefits in terms of performance improvement, but there are challenges facing the integration of D2D in LTE-A network. This section provides an in-depth discussion on these challenges.

2.4.1 Device Discovery

Device discovery is one of the major challenges in D2D communication, as devices need to identify the presence of other devices, know each other's identity, get information of services on offer, and satisfy the proximity conditions, before establishing the direct communication path. Devices can periodically broadcast their identity information so that other devices can be aware of their existence and decide whether to respond to their discovery request and subsequently initiate a D2D direct or device relaying communication. This is known as peer discovery, from D2D point of view. Peer discovery is performed by exchanging signaling messages referred to as beacon signals between users who want to communicate in D2D mode and between them and the eNB for control purposes [28]. These beacon signals contain the identity of each potential D2D user, type of service, and serves as pilot (reference) signals for measuring the channel quality indicator (CQI) of the direct path [29]. Furthermore, the devices need to determine appropriate modulation and coding schemes to be used, such as reference signals like LTE UL demodulation reference signal (DM–RS) can be inserted at the D2D Tx for channel estimation and demodulation at the D2D Rx [30]. The eNB uses the CQI value, mapped it to SINR, to establish the direct path between the D2D pair, when the SINR is above a predefined threshold, i.e. favorable for D2D communication.

There are two main techniques in D2D discovery process, namely Priori and Prosteri [31]. In former, a user initiates the discovery process by broadcasting beacon signals at regular intervals, prior to the actual communication session between the users. While in the latter, a discovery process is initiated by the eNB, while a communication session is ongoing between users. In such case, the eNB identify the users as potential D2D pairs, by analyzing their IP addresses, and therefore, recommend them to switch over to D2D mode, to achieve better performance and higher gain. In both cases, the eNB reserves a discovery resource pool, which is utilized for D2D discovery purposes. Several research works have been conducted in D2D discovery, aiming at proposing efficient peer discovery algorithms that will conform to LTE-A discovery protocols with less signaling overhead and high performance in terms of discovery speed, discovery resource utilization, energy consumption, and success probability [32]. Research is still going on to find optimal D2D discovery solutions that strike balance between cost, complexity, and performance.

2.4.2 Mode Selection

Another challenging task in D2D communication is how to select the optimal transmission mode (i.e. either cellular or D2D mode) for potential D2D users after discovering each other. Although potential D2D users might be in the vicinity of each other, but it may not be optimal for them to operate in the D2D mode from performance perspective.

Mode selection means that the network (eNB) and/or the D2D users themselves choose to operate either in D2D mode (via direct communication link) or cellular mode (via the eNB), depending on some selection metrics such as path loss [33], distance between D2D and cellular users [34], channel quality condition [35–37] interference among D2D pairs, and energy efficiency [38]. Channel quality condition and SINR is the most common selection metric proposed in the literature because it generates less signaling overhead, with less implementation complexity. The mode selection criterion is used to determine which D2D-eligible flow will actually use the direct link based on predefined SINR threshold, with the aim of maximizing system throughput and improving system capacity. On contrary, users can be in cellular mode, when there are no available resources to reuse due to interference problem, or when the D2D pair is at a distance that is not feasible for D2D communication. Table 2.2 categorized the available communication modes.

In network-controlled D2D, the eNB is responsible for selecting either D2D or cellular mode, based on the channel quality report from the users. The eNB keeps track of D2D ink quality when D2D mode is selected and handovers to cellular mode in case if the D2D proximity conditions are no longer met, or if there is a drop in the D2D link quality, while cellular link achieves higher throughput. This contrasts with autonomous D2D, where the D2D users themselves choose to operate between these modes. In essence, proper mode selection plays an important role in determining the performance of D2D communication in cellular network. This is because, when D2D mode is appropriately chosen, the reuse factor increases, while inadequate mode selection results in generating harmful interference within the system.

2.4.3 Radio Resource Management

Once the appropriate transmission mode is selected, a media access control (MAC) level packet scheduler that determines which RBs goes to which flow on

Table 2.2 User communication modes in D2D enable cellular networks.

Communication mode	Resource-sharing scheme	Spectrum efficiency
D2D-shared mode	Non-orthogonal	High
D2D-dedicated mode	Orthogonal	Low
Cellular mode	Conventional cellular resource allocation	None

Source: Safdar et al. [7].

each transmission time interval (TTI) is required. In contrast to traditional LTE scheduling, radio resources are simultaneously allocated to the D2D and cellular links. The goal is to find the optimal RB assignment solution for both the traditional cellular and D2D communication links, based on which the communication links can efficiently perform their individual transmission while avoiding undesirable interference to each other.

An RB is the basic representation of radio resources in LTE-A networks. It occupies one slot in the time domain and 180 kHz in the frequency domain, i.e. 12 subcarriers with 15-kHz subcarrier spacing [39]. The RBs are assigned using either a centralized allocation scheme or a distributed allocation scheme. The eNB is responsible for the allocation of the RBs in the centralized resource allocation scheme, while the users select from a resource pool that is preconfigured statically or semi-statically by the eNB in the distributed resource allocation scheme. Hence, eNB's participation in making scheduling decision for both the D2D and cellular users incurs more signaling and computational overhead, especially in high network density areas. The distributed resource allocation scheme is, therefore, gaining attention in D2D communication due to its scalability, low complexity, and less overhead in resource assignment. A robust spectrum sensing and access mechanism, similar to the one used in the cognitive radio networks, is highly required to reduce harmful interference to cellular users.

D2D transmissions can reuse either the uplink or downlink or both resources of the licensed cellular spectrum. UL resources are widely considered to be the efficient reuse candidates for D2D transmission because they are less utilized than DL (due to the well-known traffic asymmetry) [24, 40–42]. Also, user equipment (UE) has less transmit power compared to eNB, and the victim of D2D interference is mainly the eNB from cellular communication perspective. Therefore, in view of this, interference can easily be coordinated and controlled. However, reusing UL resources requires the mobile terminals to be equipped with single-carrier frequency-division multiple access (SC–FDMA) receiver, which results in increased system complexity than equipping the mobile terminals with orthogonal frequency-division multiple access (OFDMA) transmitter, for the case of DL resource sharing.

In addition, the same RBs can be spatially reused among different D2D pairs. This improves spectrum utilization since more UEs that are distant apart can be simultaneously served with the same RBs. However, spatial reuse generates mutual interference among the D2D users and therefore must be coordinated effectively. It is noticed that instantaneous interference among multiple D2D pair and cellular users depends on efficient resource scheduling technique [43]. Various optimization framework has been proposed in the literature, on how to jointly allocate radio resources with power control, either centralized [44], fully distributed, or semi-distributed schemes [45], to improve spectral efficiency. Thus, intelligent and reliable allocation of the shared resource blocks for D2D and cellular communications would result in optimal spectrum utilization and minimizes harmful interference in the network.

2.4.4 Modification to LTE-A Architecture

The existing centralized LTE-A architecture needs to be modified and new features have to be implemented in order to accommodate D2D procedures, such as device discovery, mode selection, D2D session management setup, physical layer procedures, and resource allocation. Raghothaman et al. [46] describes the architectural design and protocol modifications that need to be integrated on the existing cellular standard in order to support D2D communication. Similarly, the integration of new functional nodes and interfaces to the existing LTE-A architecture to support D2D services is proposed in Ref. [8], as shown in Figure 2.8.

The ProSec Function and ProSe Application Server nodes are incorporated and connected to the EPC and E-UTRAN. These nodes are responsible to aid D2D operations, such as D2D discovery, authorization and policy, device identifier allocation, call establishment procedures, mobility tracking, service identification, and other support services. On the one hand, a ProSe Application is added on the D2D user side and logically connected to ProSe Function and ProSe Application Server nodes for signaling purposes and other D2D procedure message exchange. The direct communication between D2D users is carried out via the PC5 interface. This interface is integrated in the existing physical layer design of LTE-A and will be used for exchanging all D2D control and data signals, e.g. peer discovery, synchronization, user data transfer [6]. Furthermore, the existing mobility management entity (MME) and home subscriber server (HSS) nodes are enhanced to provide user information for authorization and other D2D functionalities.

2.4.5 Security in D2D

The coexistence of the D2D and cellular communication increases the potential risk of new threats to the security of the system. This is because various combinations of

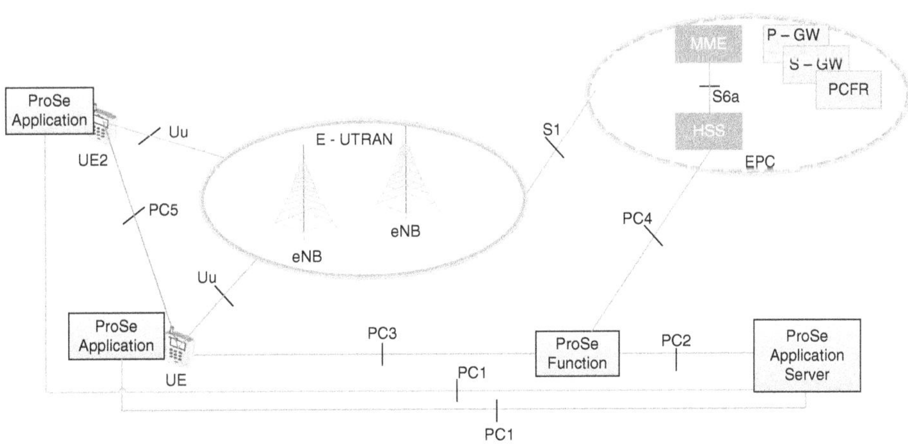

Figure 2.8 Schematic representation of D2D-enhanced LTE architecture [7].

user devices, protocols, and network topologies are integrated to work together as a single platform. This therefore makes the system more vulnerable and susceptible to different types of network attacks, such as denial of service, man-in-the-middle, and replay attacks. The security aspect of D2D is yet to be given considerate attention from the research community, and very few works is done in this area. The security framework of D2D is categorized into two types [13] namely open access and closed access. In the former, there is no restriction to any device that want to operate in D2D mode, any device can be discovered and is discoverable, and also, any device can act as relay for all other potential D2D devices. This category poses a security concern, because authentication and authorization policy is required for potential D2D users. On the other hand, the closed access provides a list of trusted devices that can be discovered and are discoverable to ensure a level of privacy. Other study is of the view that the present security framework of LTE-A network can be adopted in D2D communication, since both communications operate on the same platform [9]. Nevertheless, D2D security is an immature area which requires adequate attention in order to develop efficient cryptographic techniques that will ensure secure D2D networks with confidentiality, integrity, and availability features.

2.4.6 Mobility Management

Majority of work on D2D communication provides analysis on the single-cell deployment scenarios. Some even suggests that D2D communications should be designed for rather static nodes with limited mobility support, but still movement from one cell to another cannot be neglected. Therefore, mobility management and handover have significant impact on the performance of D2D communication. Firstly, the maximum distance between D2D pairs in different deployment scenarios in accordance with QoS requirement and interference constraints to cellular links needs to be studied. Secondly, the movement of D2D transceivers from one cell to another during an ongoing communication session is practically possible, and thus, service continuity is required. This, however, would lead to a handover. Thus, a resilience handover process is required in order to realize the seamless communication on the D2D links. Alternatively, the D2D transceivers may be switched over to cellular mode when it is no longer possible to continue transmission in D2D mode due to mobility or excessive interference levels experienced from neighboring cells. Therefore, new decision-making handover algorithms to handle movement from single- to multi-cell scenarios or to switch to cellular mode need to be proposed.

2.5 Interference in D2D

D2D communication underlaying cellular network is expected to operate within the same coverage area of an existing cell of LTE-A network and share the same cellular spectrum. Thus, reusing the same radio resource blocks of cellular users by D2D users introduces undesirable interference (known as cross-tier interference) from cellular users to D2D users and from D2D users to cellular users. When reusing the downlink RBs, D2D users suffer harmful interference from the eNB due to the high transmit power of the eNB. This makes it difficult to guarantee the quality of D2D services, decreases the SINR, and hence, results in poor performance of the D2D systems.

On the other hand, reusing uplink RBs generates less undesirable interference to the D2D users, because the traffic overhead and control signaling of uplink are much lower than that of downlink in cellular networks [32]. Hence, the total interference level in uplink spectrum is less than that in downlink spectrum.

Figures 2.9 and 2.10 illustrate the interference scenarios for UL and DL reuse cases, respectively. For UL resource sharing, it can be observed in Figure 2.9 that the D2D Tx causes undesirable interference to the eNB, while the cellular uplink user generates interference to the D2DRxs. For DL case (Figure 2.10), the eNB is the aggressor interfering with more than one D2D Rxs and D2D Tx is the aggressor interfering with cellular downlink user. Furthermore, there exists mutual interference among D2D pairs that simultaneously share the same RBs in both UL and DL reuse cases, which is referred to as co-tier interference. Therefore, it is highly necessary to mitigate interference introduced by D2D users in order not to cause service disruption to the legacy cellular users. In this research work, we are focusing on uplink resource reuse for D2D links, for better performance in terms of D2D channel rate and operability. The basic interference scenarios in D2D-enabled cellular network are summarized in Table 2.3. Cases 1 and 3 are interferences from D2D communication to legacy cellular network when reusing UL and DL resources. These interference situations have high priority in order to protect the legacy cellular users from service disruption. On the other hand, case 2 and 4 are interference situations from cellular communication to D2D users, for UL and DL reuse period. These interference cases reduce the

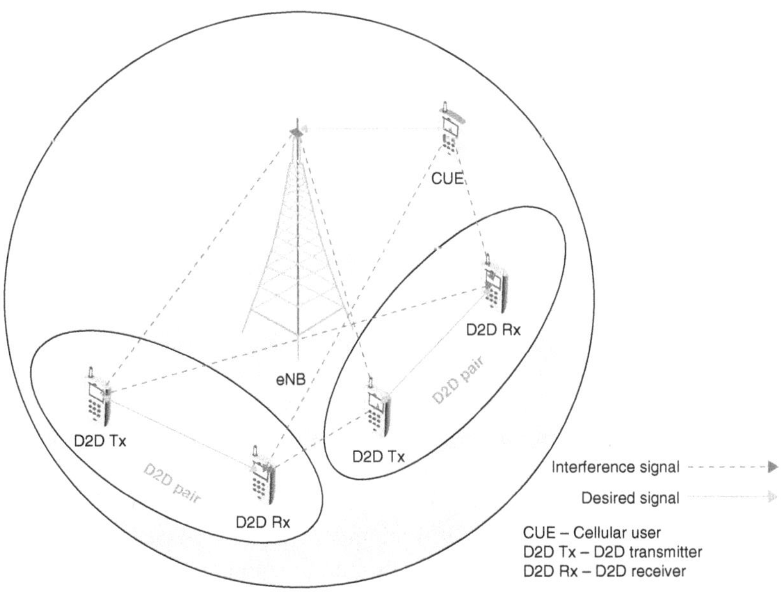

Figure 2.9 Interference scenario of D2D and cellular links under uplink resource reuse [7].

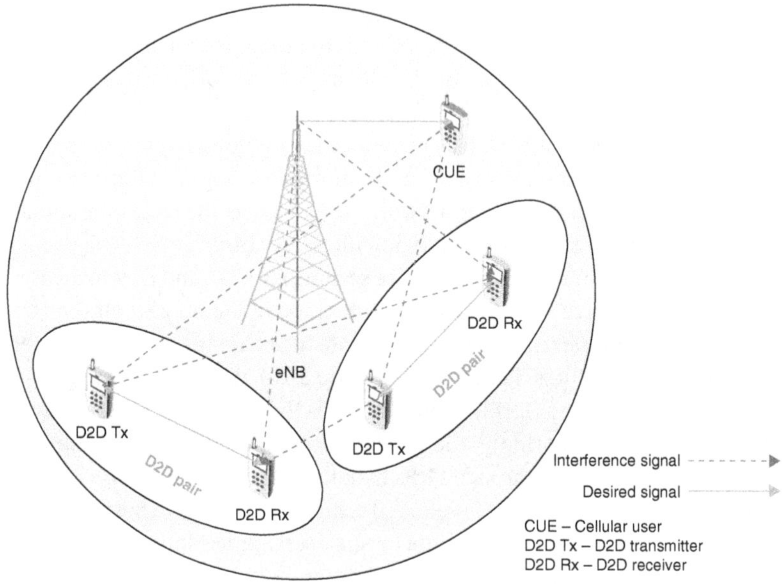

Figure 2.10 Interference scenario of D2D and cellular links under downlink resource reuse [7].

Table 2.3 Classification of interference scenarios for different resource-sharing period.

Case	Resource-sharing direction	Aggressor	Victim	Type of interference	Priority
1	UL	D2D Tx	eNB	Cross-tier	Yes
2	UL	CU	D2D Rx	Cross-tier	No
3	DL	D2D Tx	CU	Cross-tier	Yes
4	DL	eNB	D2D Rx	Cross-tier	No
5	UL/DL	D2D Tx	D2D Rx	Co-tier	No

Source: Safdar et al. [7].

reliability of D2D communication, and therefore, must be overcome. Case 5 is the interference situation among multiple D2D pairs sharing the same UL/DL resources simultaneously. This interference further degrades the performance of D2D communication. All these interference cases must be mitigated efficiently in order to guarantee the QoS of cellular network, fulfill the prioritized cellular service requirements, and improve the reliability of D2D communication.

Recently, the research community has been focusing on developing novel interference management techniques to mitigate the interference generated due to the coexistence of D2D communication in cellular networks. In traditional cellular systems, power control algorithms and RRM are often used to mitigate interference [19]. Optimal interference awareness/avoidance and coordination solutions must also be applied in D2D communication to improve the overall network throughput of both LTE-A and D2D systems and satisfy the QoS constraint.

2.5.1 Power Control Techniques

One of the most common interference avoidance schemes is adjusting the transmit power of D2D Tx below a predefined threshold while meeting the SINR target of cellular communication [15–17]. The eNB can set up constraint on the transmit power level of the D2D Tx to limit the interference caused to the cellular receivers. With adequate power control (PC) scheme, more D2D pairs can reuse the same resources simultaneously, which translate to higher spectrum efficiency. Nevertheless, this simple power control scheme results in underutilization of D2D communication among potential D2D users due to the restriction on transmission power level. In other words, D2D cannot always be feasible.

Figure 2.11 shows the schematic representation of power control strategy. The applicability of power control scheme largely depends on mutual distance between the D2D pairs, the distance between the D2D pairs, and the eNB or the cellular user equipment (CUE), for UL reuses case and DL reuse case, respectively. If the D2D pairs are far away from the eNB or CUE, and at the same time, the D2D pairs are in close proximity, reducing the power control will not affect the performance of the D2D communication. On the contrary, when the eNB or CUE are relatively close to the D2D pair and the distance between the D2D pair is large, decreasing the transmit power of D2D users could result in very low probability of D2D communication or even prevent D2D communication at all between the D2D pairs.

The power control schemes can mitigate interference from D2D to cellular communication, but they are only applicable when the D2D pairs are close to each other and/or at a sufficient distance from the eNB or CUE. Also, the stringent restriction of limiting the transmission power of D2D users may degrade the performance of D2D communication. As such, power control scheme cannot be the optimal interference mitigation solution in D2D communication.

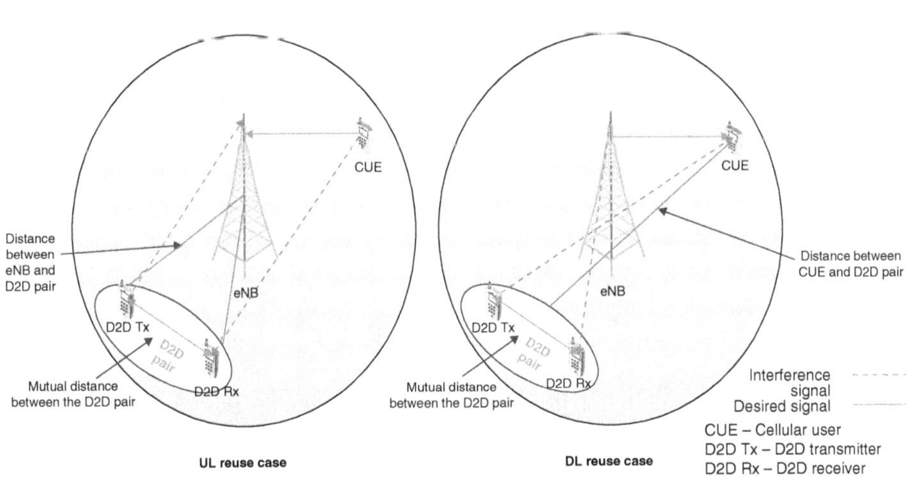

Figure 2.11 D2D power control model [7].

2.5.2 Radio Resource Allocation Techniques

A different method for interference mitigation in D2D communication is by utilizing various radio resource allocation algorithms. The main aim here is to optimally assign radio resources to a group of or all D2D pairs efficiently, and at the same time ensure that co-channel users do not interfere with each other.

A novel interference mitigation solution by intelligent selection of either UL or DL spectrum band for D2D link reuse based on the received power as the radio distance metric is proposed in Ref. [47]. The received power is measured from an eNB, if it is above a predefined threshold, then the DL is selected otherwise UL is selected for D2D link reuse. As such, the interfering signals from D2D communication to cellular users are reduced, and the overall cellular gain is improved. However, their scheme incurs more signaling overhead to the eNB and is delay bound since the band selection decision-making is carried out on every TTI.

A location-based resource allocation method to mitigate mutual interference between D2D and CU users sharing the same UL resources was studied in Ref. [25] through the concept of accessible and reusable regions. Only D2D users located in the accessible regions can simultaneously share the same radio resources with cellular users that are in reusable regions. Otherwise, the reuse scheme is not feasible. From their results, the outage probabilities of both D2D and cellular users are minimized, with improved system reliability. However, localization of users bounded by regional areas yields less spectrum efficiency and flexibility. Similarly, a distance-based resource allocation scheme to mitigate interference from CUE to D2D Rx is proposed in Ref. [48]. This scheme benefits from low signaling overhead, because the eNB assigns resources to the D2D pair according to the mutual distance between them and CU, rather than CSI measurement between the links. Though, the exact location of individual cellular and D2D users need to be known by the eNB in order to share the resources effectively.

The problem of interference management in D2D communication underlaying LTE-A network is formulated as an interference-aware graph model [39]. An interference-aware graph is then proposed based on the radio resource reuse algorithms that can effectively improve the system throughput and mitigate co-channel interference among D2D and cellular users. Presented simulation results demonstrate an improved system performance in terms of overall network sum rate, with low computational overhead. Also, Ref. [49] has modeled the problem of co-channel interference between D2D and cellular users with the aid of an advanced mathematical tool, game theory. In their work, an interference-aware resource allocation using sequential second price auction scheme was proposed to optimize the overall sum rate of the D2D system. Simulation results demonstrate an achievable performance of the system sum rate, but at high complexity cost.

A mechanism to avoid inter-cell near–far interference problem in a multi-cell environment is proposed in Ref. [50]. In their procedure, the neighboring eNBs monitor the control channels of a D2D subsystem and exchange necessary information, to identify the interference D2D victims and CUs causing the interference due to UL spectrum sharing. Based on this knowledge, the serving eNB can stop scheduling transmission of interferer (i.e. CU) until D2D transmission ends. Simulation results shows that the performance of D2D communication is enhanced, but at the cost of reliable control channel sensing. A similar procedure was adopted in Ref. [49] to solve the interference problem in a single-cell scenario.

An interference mitigation solution based on evaluating the neighbor distance and Tolerant Interference Degree (TID) level among potential D2D pairs is proposed in [51]. The TID performance metric was defined as the number of D2D pairs that can coexist with a given D2D pair to reduce the undesirable interference at the D2D Rx. Although their approach is less complex and incurs less signaling overhead to the eNB, but the orthogonal resource allocation scheme considered is less spectrum efficient.

The objective of radio resource allocation techniques is to optimize the resource usage between primary cellular users and D2D pairs. Adopting these diverse techniques suppress the interference problem when D2D communication coexists with cellular network. However, the main drawback with this approach is underutilization of radio resources and reducing multi-user diversity, because the physical separation limits the scheduling alternatives of the eNB.

2.5.3 Joint Power Control and Radio Resource Allocation Techniques

A more advanced approach for mitigating interference in D2D/LTE system is to jointly use power control with various resource allocation techniques in order to realize the full potentials of D2D communication. Several works have investigated this joint optimization problem, where they focus on interference management and control between cellular links and D2D links, with the aim of complementing and enhancing the overall performance of a single scheme.

A combined effect of dynamic power control and resource allocation to reduce D2D interference to cellular network was studied in Ref. [18]. In their approach, the eNB initially assigns resources to CUE, then to D2D users and finally reuses CUE's resources to D2D users when the demand is high. Then, the eNB dynamically adjusts the transmit power of the D2D Tx by determining the channel gain between individual terminals in order to avoid harmful interference when both D2D and CUE occupy the same resources. The performance of this scheme was measured based on the SINR level achieved in both transmission systems. However, the centralized nature of dynamic power control by eNB incurs significant overhead on part of the eNB.

The authors describe a power control and distance-based resource allocation mechanism to mitigate the interference among cellular and D2D users sharing UL resources [1]. To avoid interference from cellular users to D2D communication, the D2D users only reuses the UL resource blocks of CUs that are not in close proximity. On the other hand, the D2D users adjust their transmission power in a manner that the interference from D2D communication to eNB is below a tolerable threshold. Their results reveal that the overall system throughput is enhanced by 41% under the proposed mechanism.

Meanwhile, a novel interference avoidance model based on user location is proposed in Ref. [52]. In their work, rather than limiting the transmit power of D2D users as in conventional interference management scheme, an interference-limited area is proposed. Within this area, no cellular user will share the same resources with a D2D pair. As such, excessive interference between D2D and cellular communication is avoided. Although their simulation results show a significant performance gain, however, the major drawback of this scheme is reducing multi-user diversity because the physical constraint limits the scheduling efficiency of the eNB.

A power control and distance-based interference mitigation algorithms have been proposed in Ref. [53]. The scheme limits the maximum transmit power of the D2DTx in accordance with the minimum acceptable SINR target of the eNB, such that the interfering signal from the D2DTx to eNB is reduced. Then, the eNB selects the optimal UL resources to share with the D2D link, by estimating the distances between D2D users and various CUEs using location estimator. Then, the CU that minimizes the outage probability of the D2D link is chosen accordingly. In other words, the longer the distance between the CUE and D2D link, the better performance D2D communication would experience. As such, the interference from CU to D2DRx is reduced. Their numerical results show that the performance of D2D communication is improved in terms of outage probability gain. However, mutual interference among D2D pairs sharing the same radio resources was ignored.

2.5.4 Spectrum Splitting Techniques

Spectrum splitting is the easiest way to avoid interference in D2D-enabled cellular network. Adopting time division multiplexing (TDM) technology to separate cellular and D2D transmissions could effectively reduce the interference level in the hybrid cellular network with D2D communication [54–56]. However, this method would lead to inefficient utilization of the available spectrum. Also, it only accounts for cross-tier interference between cellular and D2D users. Therefore, additional mechanism is required to mitigate the interference among multiple D2D users, which share the same set of resources.

2.5.5 Other Interference Mitigation Techniques

A solution to interference issue in D2D communication through the application of network coding technique has been proposed in Ref. [57]. To mitigate the interfering D2D signals to eNB, helper nodes are selected to assist in cellular UL transmission to eNB, and network coding is applied for the actual transmission. Although the performance of cellular communication is improved, but on the other hand, additional interference is generated by the helper nodes, which have negative influence on D2D transmission.

The application and impact of interference-aware interference mitigation algorithms in network-assisted D2D communication are investigated in Ref. [27]. In the proposed scheme, the gain of interference-aware algorithms are evaluated using simultaneous non-unique decoding (SND), and decoding cellular interfering signals at the D2D Rx reduces interference. Their results indicate that interference-aware algorithms enhance the throughput of D2D communication. However, both D2D and cellular users need to operate on interference-aware algorithms under the control of an eNB.

From the foregoing, it can be observed that majority of the existing literatures focus extensively on proposing interference mitigation solutions based on transmission power control of D2D Txs, diverse resource allocation techniques, interference-limited areas, and combined power control and resource allocation mechanisms, etc.

2.5.6 Multiple-Input Multiple-Output Techniques

MIMO antenna systems have become an important component in today's cellular wireless network standards to improve the overall system performance [58]. These MIMO transmission methods such as beamforming and interference cancelation technique can be utilized in D2D communication underlaying cellular communication to further avoid interference between cellular links and D2D links with the prior knowledge of the interfering channels of CSI.

Beamforming also utilizes multiple transmit and receive antenna elements to generate directional antenna beam pattern. The generated beam (desired signal) is steered in the direction of the intended recipient, while at the same time, cancelling out undesirable interference in the direction of other users [59]. In other words, the transmit power is maximized toward the receiver angle, while minimizing the signal in the null space. When transmitting, a beamformer controls the phase and relative amplitude of the signal at each transmitter, thereby producing a high directional beam pattern in the direction of the intended recipient and null in the direction of interference. This increases the SINR of the intended user and reduces the wastage of transmitted power in the undesired direction. Other benefits of beamforming include high spectrum reuse factor, increase in capacity, etc. Therefore, the use of such multi-antenna beamforming either at downlink or uplink transmission can mitigate the interference levels between cellular and D2D transmission, improves system capacity, and further guarantees the feasibility of D2D transmission.

eNB beamforming that is performing beamforming on the cellular downlink (eNB) to mitigate the effect of interference has gained attention in the research community and some literatures exist in this area. In such a system design approach, the eNB is equipped with an array of multiple antennas with different precoding strategies. In particular, the eNB avoid generating cross-tier interference to D2D Rx (UE3) sharing the same resources by aligning the transmitted signal from the eNB to the null space of the eNB–D2D interference channel, as illustrated in Figure 2.12.

Investigations have been made in Ref. [60] about the performance of D2D communication system sharing downlink resources with multi-antenna eNB, for both beamforming and interference cancelation strategies at the eNB under quantized channel estimation and perfect CSI. Also, D2D receivers' interference mitigation schemes are proposed in Ref. [17] which use MIMO eNB downlink transmission, in a

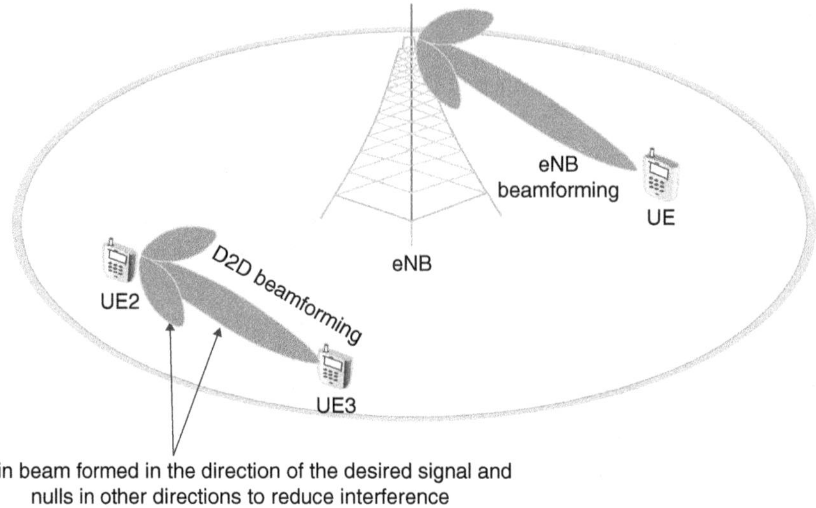

Figure 2.12 Schematic representation of beamforming with eNB and D2D users equipped with multiple antenna elements [7].

single-cell scenario. The eNB can utilize any MIMO transmission scheme by designing transmitter weights for a projected downlink channel, and then, the downlink-precoding matrix can be computed as the multiplication product of the projection matrix and the designed transmitter weights. The simulation results have shown that D2D links experiences higher SINR thereby increasing the reuse factor, whereas the CUE undergoes marginal decrease in SINR. However, this approach only works when the downlink RBs are being considered as the reuse resources for D2D transmission. Moreover, the eNB requires the CSI of the interfering links.

Performance-related investigation of a joint beamforming and power control method to mitigate the interference from eNB to D2D users in the downlink resource reuse scenario is made in Ref. [19]. In their scheme, the eNB is equipped with multiple antennas and performs beamforming to avoid the interference experienced by D2D Rxs, while the user terminals have single antennas. The eNB calculates the beamforming matrix based on the interfering link CSI obtained from D2D Rxs and data link CSI obtained from the cellular user. The eNB then determines transmit powers based on the SINR thresholds of both cellular and D2D links. The presented results show that beamforming improves the performance of D2D communication such that the SINR criteria limit the interference experienced by D2D Rxs from eNB. Also, controlling the transmit power of the D2D Tx enhances the performance of the cellular communication, as it reduces the interference experienced by cellular users. However, this scheme is based on the downlink resource sharing, in which eNB has high transmit power that subjects the D2D Rxs to more excessive interference signals. In addition, their analysis is based on single-cell deployment, without considering interference from neighboring cells.

An optimized joint beamforming with power control scheme to reduce the mutual interference that coexist between D2D and cellular users and to minimize power consumption while satisfying the QoS constraint of both systems is proposed in Ref. [20]. The optimization problem for the transmit power and beamforming weight vectors was solved based on support vector machine (SVM) algorithms, with the aid of statistical CSI estimation. A novel analytical expression of the ergodic capacity (EC) and average symbol error rate (ASER) of all the users in the system was obtained. Both system simulation results and theoretical analysis have shown that the proposed scheme achieves a good performance in terms of system throughput and capacity. However, their performance analysis considered only single D2D pair within the system model. Thus, multiple D2D pairs are needed to be deployed in order to ascertain the level of interference suppression of their proposed scheme. An interference-aware scheduling algorithm for D2D communication in multi-antenna system is proposed in Ref. [37]. In their approach, D2D links are only paired to share the same UL resources with a CU such that the interference caused to both communication systems are below certain thresholds; otherwise, the D2D links are scheduled on default resources. Then, an optimum beamforming is applied based on SINR metric to further reduce the interference level and improve the system performance. Simulation results show that the pairing algorithm together with beamforming technique increases overall system throughput while satisfying QoS constraints of both D2D and cellular communication.

Few studies have exploited multi-antenna beamforming technique on the DL eNB infrastructure to mitigate interference experienced by D2D Rxs. On the other hand, D2D beamforming, that is, performing beamforming on user devices to avoid any undesirable interference generated from D2D Tx to cellular user and to other co-channel D2D users, is still yet to be studied extensively. To support this approach, already, uplink multi-antenna transmission is among add-on features proposed in Release 10 of LTE-A [19], as part of 3GPP's effort to satisfy the requirements of the LTE-A system.

Therefore, uplink multi-antenna design on user devices can be utilized to achieve D2D beamforming, by steering the beam toward D2D Rx and null in other direction. This will effectively reduce the interference levels in D2D communication. But, to the best of our knowledge, the only work that exploits the D2D beamforming in which a null-space-based robust interference mitigation scheme for D2D systems sharing UL cellular resources is modeled is proposed in Refs. [61, 62]. In their work, the interfering channel's CSI to eNB and from CU was estimated using linear minimum mean square error (LMMSE) method. Then, transmit and receive beamforming were designed at the D2D transceivers pointing signals only to the direction of the null space estimated channel in order to effectively minimize the interference generated to eNB and from CU. Simulation results shows that the scheme improves D2D system throughput. However, there was not any comparison with other similar studies to indicate the relevance of their optimized interference technique [63–65].

Table 2.4 shows a summary of the studies on interference management for D2D-enabled cellular networks in terms of interference mitigation technique, D2D reuse case, interference mitigation case, and scenario. It can be observed that the techniques of interference mitigation differ slightly, depending on the interference scenario to be solved. For instance, the power control technique is

Table 2.4 Summary of literatures on interference mitigation techniques in D2D communication underlaying LTE-A network.

Article	Interference mitigation technique	D2D reuse resources	Interference type	Network complexity
[1, 18]	Joint power control with resource allocation	UL	Cellular to D2D and vice versa	Single cell
[17, 20, 21]	Joint power control with MIMO beamforming	DL	Cellular to D2D	Single cell
[15–17]	Power control with SNIR target	DL/UL	D2D to cellular	Single cell
[25, 47, 48, 53]	Distance/location-based resource allocation	UL	Cellular to D2D and vice versa	Single cell
[27]	Interference-aware algorithms	UL/DL	Cellular to D2D	Single cell
[54]	Time division duplexing	UL/DL	Cellular to D2D	Single cell
[51]	Greedy orthogonal resource allocation	UL/DL	Between multiple D2D users	Single cell
[57, 62]	Network coding	UL	D2D to cellular	Single cell
[60]	Beamforming and interference cancelation	DL	Cellular to D2D	Single cell
[61]	MIMO beamforming	UL	D2D to cellular	Single cell
[50]	Channel-based resource allocation	UL	Cellular to D2D	Multi cell

Source: Safdar et al. [7].

utilized mostly for interference from D2D to cellular communication. On the other hand, diverse radio resource allocation and advanced antenna techniques such as MIMO, beamforming addresses both cross-tier and co-tier interference problems. Also, majority of the previous works considers one D2D pair scenario in their system model. Hence, multiple D2D pair case needs to be investigated to mimic practical cellular network environment. With multiple D2D pair, the reuse factor increases, which results in increase in resource utilization.

However, this increases the interference levels, which requires to be managed effectively, while meeting the QoS requirement of both D2D and CU users. Also, 99% of studies on D2D and their performance analysis are based on interference within a single-cell scenario. However, this assumption is far from reality, as interference from neighboring cells including the interference from both cellular and D2D transmissions in the neighboring cells should be taking into consideration, as illustrated in Figure 2.13. In fact, the performance of cell edge users is predominantly affected by neighboring cell interferences.

In Figure 2.13, although the D2D pair (i.e. D2D Tx and D2D Rx) is in close proximity, they are two boundary users located in different cells, cell 1 and cell 2, respectively. Sharing the same resources implies that they may be interfered by several cellular users from neighboring cells, such as CUE 11 in Cell 1, CUE 21 in Cell 2,

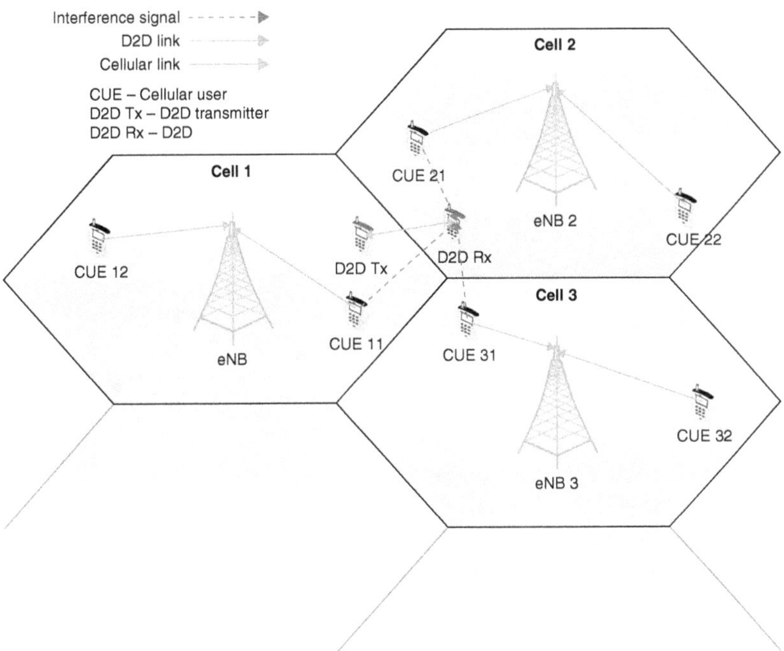

Figure 2.13 Interference scenario in multi-cell network [7].

and CUE 31 in Cell 3. The interference mitigation solutions proposed for single-cell system cannot be equally applied in a multi-cell environment due to the independent RRM and coordination by each eNB. This means that the serving eNBs in each cell most coordinate in allocating resources for D2D reuse, thereby mitigating the inter-cell interference problem.

2.5.7 Comparative Analysis of D2D Interference Mitigation Techniques

Most of the interference mitigation techniques for D2D-enabled cellular network in the literature employs the PC, radio resource allocation (RRA), hybrid/joint, and multi-antenna schemes (Figure 2.14). Other schemes proposed such as networking coding, spectrum splitting, etc., were not given considerate attention. Hence, comparative analysis of the main schemes, i.e. PC, RRA, and MIMO, is already presented in Table 2.5, whereas Table 2.6 shows pros and cons of the above-mentioned interference mitigation schemes. To sum it all, techniques to minimize the total power consumption are simple to implement but offer limited performance gain from D2D communication perspective.

Majority of the RRA techniques employed (i) channel sensing and (ii) geographical user location-aware reuse approaches to schedule cellular and D2D users efficiently, while avoiding interference in the cross-tier system. Although these RRA schemes are viable solution to effectively mitigate interference, but these methods mostly do not admit D2D pairs to operate on frequency resources that will violate the required QoS constraints of the cellular users. This means that the D2D pair may

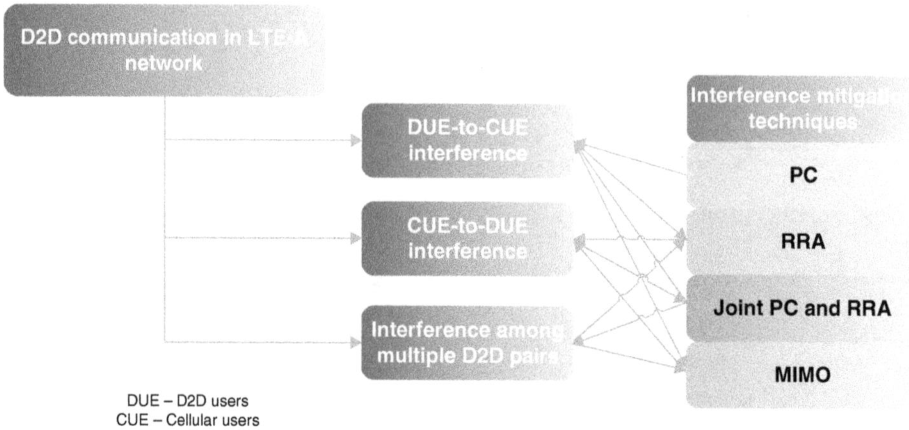

Figure 2.14 Overview of interference mitigation techniques in D2D communication underlaying LTE-A network [7].

Table 2.5 Comparative analysis of interference mitigation techniques in D2D communication.

	Power control		Radio resource allocation		MIMO
Characteristics	**Fixed**	**Dynamic**	**Channel-based RA**	**Distance-based RA**	**Beamforming**
Target	Cellular QoS guaranteed, maximized throughput	Cellular QoS guaranteed, maximized throughput	Interference reduction, D2D capacity enhancement, improved SINR	Interference reduction, D2D capacity enhancement, improved SINR	Improved D2D link quality, interference avoidance, maximized total system throughput
Central control	eNB	eNB	Depends on design requirement	Depends on design requirement	Depends on design requirement
Complexity/cost	Low	High	High	Medium	High
Interference reduction level	Low	Low	High	Medium	High
Interference control type	Centralized	Centralized	Centralized	Centralized	Centralized/distributed
Spectrum efficiency	Medium	Medium	Low	Low	High
Side information	Control signals	Periodical CSI feedback	Channel measurement and feedback	User location information via GPS	CSI
Flexibility	Fixed	Dynamic	Dynamic		

Source: Safdar et al. [7].

Table 2.6 Pros and cons of D2D interference mitigation techniques.

Scheme	Pros	Cons
Power control	• Transmission power can be optimized in an adaptive manner to suppress interference • Simple to implement, especially when the D2D pairs are in close range • Low computational cost	• Low probability of D2D communication between D2D pairs due to limited transmit power • Not effective in mitigating interference from cellular users to D2D users • Limited performance from D2D communication perspective • Unable to dynamically reflect channel variations, especially when fixed PC is employed
RRA	• Resource allocation algorithms can be optimized to suitably avoid interference • Fractional frequency reuse improves channel quality by utilizing different resources for D2D and cellular communication	• Inefficient utilization of the licensed spectrum • Requires channel statistics and exact knowledge of user location, which incurs additional signaling overhead to the eNB • Longer scheduling time due to large signaling overhead • Low throughput for high payload case • Computationally intensive when multiple D2D pairs share the same resources with a cellular user
Joint PC and RRA	• Combines the advantages of PC and RRA	• Computationally intensive • Requires implementation of joint RRA and power algorithms • Increases coordination and signaling overhead
Beamforming	• Robust interference mitigation solution • Directional beam targets the intended D2D recipient while creating a null toward other users	• Requires precoder design, which incurs computational overhead • Requires accurate CSI of all involved links • Cost of hardware (multiple antenna elements) implementation

Source: Safdar et al. [7].

not operate on some radio resources at all, which translate to inefficient resource utilization. In addition, the RRA techniques are mostly based on one-to-one matching policy, in which only one D2D pair reuses the cellular resources either in the UL or DL period. This also results in low-frequency reuse gain. Joint PA and RRA solutions have high implementation cost because of the complexity of the proposed joint algorithms. It is therefore necessary to find a trade-off between high system performance and low complex algorithm.

The common characteristics of PC, RRA, and the joint/hybrid interference mitigation techniques in D2D communication are that they involve mode selection criterion, which implies that D2D communication cannot be feasible on some radio resources, even if the D2D pair is in close proximity [66, 67]. On the other hand,

MIMO techniques are more promising and yield remarkable performance compared with other interference mitigation techniques. However, these MIMO schemes are yet to be exploited in the UL reuse direction. As earlier mentioned, reusing UL channel may have a better performance than reusing DL channel.

2.6 Summary

D2D communication is a promising technology that aims at maximizing system throughput through enhanced spectrum efficiency. However, spectrum reuse results in harmful interference among the primary cellular users and the secondary D2D users, in addition to mutual interference between multiple D2D pairs that are sharing the same resources. This undesirable interference degrades the overall network performance, which must be tackled. Various interference mitigation schemes such as power control, efficient resource allocation, and multi-antenna beamforming among others have been reviewed and critically analyzed. The power control scheme is not enough to handle the mutual interference between D2D communication and cellular network. Various resource allocation strategies proposed which are aimed at eliminating interference problem in D2D/LTE system lead to underutilization of the licensed spectrum. Furthermore, the multi-antenna beamforming schemes studied mainly focused on suppressing downlink interference from eNB to D2D Rxs while ignoring the problem of uplink precoding for interference mitigation from D2D communication to cellular users. This necessitates the need to further investigate this area in order to efficiently mitigate cross-tier and co-tier interference during uplink resource sharing. This will guarantee the performance of the cellular network, improve D2D link quality, and enhance the reliability of D2D communication.

References

1 Peng, T., Lu, Q., Wang, H., Xu, S., and Wang, W. (2009). Interference avoidance mechanisms in the hybrid cellular and device-to-device systems. *Proceedings of the IEEE International Symposium on Personal, Indoor and Mobile Radio Communications*.

2 Abbasi, Q.H., Ur-Rehman, M., Qaraqe, K., and Alomainy, A. (ed.) (2016). Advances in body-centric wireless communication: applications and state-of-the-art. *The IET*.

3 Mustafa, H.A., Shakir, M.Z., Imran, M.A. et al. (2015). Coverage gain and device-to-device user density: stochastic geometry modeling and analysis. *IEEE Communications Letters* 19 (10): 1742–1745.

4 Yue, Z., Sun, W., Li, P., Ur-Rehman, M., and Yang, X. (2015). Internet of things: Architecture, technology and key problems in implementation. *Proceedings of the International Congress on Image and Signal Processing (CISP)*, 1298–1302.

5 Mustafa, H.A., Imran, M.A., Shakir, M.Z. et al. (2016). Separation framework: an enabler for cooperative and D2D communication for future 5G networks. *IEEE Communication Surveys and Tutorials* 18 (1): 419–445, First quarter 2016.

6 Lin, X., Andrews, J., Ghosh, A., and Ratasuk, R. (2014). An overview of 3GPP device-to-device proximity services. *IEEE Communications Magazine* 52 (4): 40–48.

7 Safdar, G.A., Ur-Rehman, M., Muhammad, M. et al. (2016). Interference mitigation in D2D communication underlaying LTE-A network. *IEEE Access* 4: 7967–7987.

8 3GPPb (2014). Study on LTE Device to Device Proximity Services: Radio Aspects. *TS 36.843*, v12.0.1, March 2014.

9 Alam, M., Yang, D., Rodriguez, J., and Abd-alhameed, R. (2014). Secure device-to-device communication in LTE-A. *IEEE Communications Magazine* 52 (4): 66–73.

10 Qiao, J., Shen, X., Mark, J. et al. (2015). Enabling device-to-device communications in millimeter-wave 5G cellular networks. *IEEE Communications Magazine* 53 (1): 209–215.

11 Zhang, B., Li, Y., Jin, D. et al. (2015). Social-aware peer discovery for D2D communications underlaying cellular networks. *IEEE Transactions on Wireless Communications* 14 (5): 2426–2439.

12 Cho, B., Koufos, K., Jantti, R., Li, Z., and Uusitalo, M. (2015). Spectrum allocation for multi-operator device-to-device communication. *Proceedings of the IEEE International Conference on Communications (ICC).*

13 Jo, M., Maksymyuk, T., Strykhalyuk, B., and Cho, C. (2015). Device-to-device-based heterogeneous radio access network architecture for mobile cloud computing. *IEEE Wireless Communications* 22 (3): 50–58.

14 Ur-Rehman, M., Chen, X., Parini, C.G., and Ying, Z. (2012). Evaluation of a statistical model for the characterization of multipath affecting mobile terminal GPS antennas in sub-urban areas. *IEEE Transactions on Antennas and Propagation* 60 (2): 1084–1094.

15 Khoueiry, B. and Soleymani, M. (2015). A novel coding strategy for device-to-device communications. *Proceedings of the IEEE Consumer Communications and Networking Conference (CCNC).*

16 Doppler, K., Rinne, M., Wijting, C. et al. (2009). Device-to-device communication as an underlay to LTE-advanced networks. *IEEE Communications Magazine* 47 (12): 42–49.

17 Araújo, D.C., Maksymyuk, T., de Almeida, A.L.F. et al. (2016). Massive MIMO: survey and future research topics. *IET Communications* 10 (15): 1938–1946.

18 Gu, J., Bae, S.J., Choi, B., and Chung, M.Y. (2011). Dynamic power control mechanism for interference coordination of device-to-device communication in cellular networks. *Proceedings of the International Conference on Ubiquitous and Future Networks (ICUFN).*

19 Song, L., Niyato, D., Han, Z., and Hossain, E. (2015). *Wireless Device-to-Device Communications and Networks*. Cambridge University Press.

20 Huang, L., Su, Z., Gao, Z. et al. (2014). OOP-based device-to-device communication simulator design of LTE network. *Proceedings of the International Conference on Computer Science & Education.*

21 Mach, P., Becvar, Z., and Vanek, T. (2015). In-band device-to-device communication in OFDMA cellular networks: a survey and challenges. *IEEE Communication Surveys and Tutorials* 17 (4): 1885–1922.

22 Zhu, D., Wang, J., Swindlehurst, A., and Zhao, C. (2014). Downlink resource reuse for device-to-device communications underlaying cellular networks. *IEEE Signal Processing Letters* 21 (5): 531–534.

23 Zhang, X., Cui, Q., Bai, J., and Tao, X. (2015). Intra-cell and inter-cell interference-constrained D2D communication underlaying cellular networks. *Electronics Letters* 51 (14): 1117–1119.

24 Byungjin, C., Koufos, K., and Jantti, R. (2014). Spectrum allocation and mode selection for overlay D2D using carrier sensing threshold. *Proceedings of the International Conference on Cognitive Radio Oriented Wireless Networks and Communications (CROWNCOM)*, 26–31.

25 Bao, P., Yu, G., and Yin, R. (2013). Novel frequency reusing scheme for interference mitigation in D2D uplink underlaying networks. *Proceedings of the International Wireless Communications and Mobile Computing Conference (IWCMC).*

26 Asadi, A., Qing, W., and Mancuso, V. (2014). A survey on device-to-device communication in cellular networks. *IEEE Communication Surveys and Tutorials* 16 (4): 1801–1819.

27 Kwon, H., Lee, J., and Kang, I. (2014). Interference-aware interference mitigation for device-to-device communications. *Proceedings of the IEEE Vehicular Technology Conference (VTC).*

28 Simsek, M., Merwaday, A., Correal, N., and Guvenc, I. (2013). Device-to-device discovery based on 3GPP system level simulations. *Proceedings of the IEEE GLOBECOM.*

29 Hong, J., Park, S., Kim, H. et al. (2013). Analysis of device-to-device discovery and link setup in LTE networks. *Proceedings of the IEEE International Symposium on Personal, Indoor, and Mobile Radio Communications (PIMRC).*

30 Zou, K., Wang, M., Yang, K. et al. (2014). Proximity discovery for device-to-device communications over a cellular network. *IEEE Communications Magazine* 52 (6): 98–107.

31 Usman, M., Gebremariam, A., Raza, U., and Granelli, F. (2015). A software-defined device-to-device communication architecture for public safety applications in 5G networks. *IEEE Access* 3: 1649–1654.

32 Zihan, E., Choi, K., and Kim, D. (2015). Distributed random access scheme for collision avoidance in cellular device-to-device communication. *IEEE Transactions on Wireless Communications* 14 (7): 3571–3585.

33 Xing, H. and Hakola, S. (2010). The investigation of power control schemes for a device-to-device communication integrated into OFDMA cellular system. *Proceedings of the IEEE International Symposium on Personal, Indoor and Mobile Radio Communications (PIMRC).*

34 ElSawy, H., Hossain, E., and Alouini, M. (2014). Analytical modeling of mode selection and power control for underlay D2D communication in cellular networks. *IEEE Transactions on Communications* 62 (11): 4147–4161.

35 Yu, C.H., Doppler, K., Ribeiro, C., and Tirkkonen, O. (2011). Resource sharing optimization for device-to-device communication underlaying cellular networks. *IEEE Transactions on Wireless Communications* 10 (8): 2752–2763.

36 Wang, B., Chen, L., Chen, X. et al. (2011). Resource allocation optimization for device-to-device communication underlaying cellular networks. *Proceedings of the IEEE Vehicular Technology Conference (VTC).*

37 Han, M., Kim, B., and Lee, J. (2012). Subchannel and transmission mode scheduling for D2D communication in OFDMA networks. *Proceedings of the IEEE Vehicular Technology Conference (VTC).*

38 Akkarajitsakul, K., Phunchongharn, P., Hossain, E., and Bhargava, V. (2012). Mode selection for energy-efficient D2D communications in LTE-advanced networks: a coalitional game approach. *Proceedings of the IEEE International Conference on Communication Systems (ICCS).*

39 Zhang, R., Cheng, X., Yang, L., and Jiao, B. (2013). Interference-aware graph based resource sharing for device-to-device communications underlaying cellular networks. *Proceedings of the IEEE Wireless Communications and Networking Conference (WCNC).*

40 Tsolkas, D., Liotou, E., Passas, N., and Merakos, L. (2014). LTE-A access, core, and protocol architecture for D2D communication. In: *Smart Device to Smart Device Communication* (ed. S. Mumtaz and J. Rodriguez), 23–40. Springer.

41 Sheng, M., Sun, H., Wang, X. et al. (July 2015). On-demand scheduling: achieving QoS differentiation for D2D communications. *IEEE Communications Magazine* 53 (7): 162–170.

42 Kpojime, H.O. and Safdar, G.A. (2015). Interference mitigation in cognitive-radio-based femtocells. *IEEE Communication Surveys and Tutorials* 17 (3): 1511–1534, third quarter 2015.

43 Han, H., Zhu, C., Viorel, D., and Ito, A. (2013). Resource allocation and beamforming algorithm based on interference avoidance approach for device-to-device communication underlaying LTE cellular network. *Communications and Network* 5 (3): 367–373.

44 Wang, F., Xu, C., Song, L., and Han, Z. (2015). Energy-efficient resource allocation for device-to-device underlay communication. *IEEE Transactions on Wireless Communications* 14 (4): 2082–2092.

45 Vlachos, C. and Friderikos, V. (2014). Robust randomized resource allocation for device-to-device communications. *Proceedings of the IEEE International Workshop on Computer Aided Modeling and Design of Communication Links and Networks (CAMAD).*

46 Raghothaman, B., Deng, E., Pragada, R. et al. (2013). Architecture and protocols for LTE-based device to device communication. *Proceedings of the International Conference on Computing, Networking and Communications (ICNC).*

47 Silva, E.C., Batista, R., da Silva, J. et al. (2015). Interference mitigation using band selection for network-assisted D2D communications. *Proceedings of the IEEE Vehicular Technology Conference (VTC).*

48 Duong, Q., Shin, Y., and Shin, O. (2013). Resource allocation scheme for device-to-device communications underlaying cellular networks. *Proceedings of*

the International Conference on Computing, Management and Telecommunications (ComManTel).

49 Sun, H., Wildemeersch, M., Sheng, M., and Quek, T.Q.S. (2015). D2D enhanced heterogeneous cellular networks with dynamic TDD. *IEEE Transactions on Wireless Communications* 14 (8): 4204–4218.

50 Xu, S., Wang, H., and Chen, T. (2012). Effective interference cancellation mechanisms for D2D communication in multi-cell cellular networks. *Proceedings of the IEEE Vehicular Technology Conference (VTC).*

51 Xu, Y., Liu, Y., Yang, K. et al. (2013). Interference mitigation scheme for device-to-device communication with QoS constraint. *Proceedings of the IEEE International Symposium on Personal, Indoor, and Mobile Radio Communications (PIMRC).*

52 Lin, Z., Gao, Z., Huang, L. et al. (2015). Hybrid architecture performance analysis for device-to-device communication in 5G cellular network. *Mobile Networks and Applications* 20 (6): 713–724.

53 Duong, Q. and Shin, O.S. (2013). Distance-based interference coordination for device-to-device communications in cellular networks. *Proceedings of the International Conference on Ubiquitous and Future Networks (ICUFN).*

54 Qin, Y., Ding, M., Zhang, M. et al. (2014). Relaying robust beamforming for device-to-device communication with channel uncertainty. *IEEE Communications Letters* 18 (10): 1859–1862.

55 Mustafa, H.A., Shakir, M.Z., Imran, M.A., and Tafazolli, R. (2015). Distance based cooperation region for D2D pair. *Proceedings of the IEEE Vehicular Technology Conference (VTC)*, 1–6.

56 Mustafa, H.A., Shakir, M.Z., Sambo, Y. et al. (2014). Spectral efficiency improvements in HetNets by exploiting device-to-device communications. *Proceedings of the IEEE GLOBECOM*, 857–862.

57 Wang, L., Tian, F., Svensson, T. et al. (2015). Exploiting full duplex for device-to-device communications in heterogeneous networks. *IEEE Communications Magazine* 53 (5): 146–152.

58 Ge, X., Zi, R., Wang, H. et al. (2016). Multi-user massive MIMO communication systems based on irregular antenna arrays. *IEEE Transactions on Wireless Communications* 15 (8): 5287–5301.

59 Razavizadeh, S., Ahn, M., and Lee, I. (2014). Three-dimensional beamforming: a new enabling technology for 5G wireless networks. *IEEE Signal Processing Magazine* 31 (6): 94–101.

60 Ni, Y., Jin, S., Xu, W. et al. (2016). Beamforming and interference cancellation for D2D communication underlaying cellular networks. *IEEE Transactions on Communications* 64 (2): 832–846.

61 Fu, W., Yao, R., Gao, F. et al. (2013). Robust null-space based interference avoiding scheme for D2D communication underlaying cellular networks. *Proceedings of the IEEE Wireless Communications and Networking Conference (WCNC).*

62 Wang, S., Hou, R., Lui, K. et al. (2015). A novel interference management scheme in underlay D2D communication. *Proceedings of the IEEE Vehicular Technology Conference (VTC).*

63 Amin, B., Ramadan, Y., Ibrahim, A., and Ismail, M. (2015). Power allocation for device-to-device communication underlaying massive MIMO multicasting networks. *Proceedings of the IEEE Wireless Communications and Networking Conference (WCNC).*

64 Goakar, R. and Cheeran, A. (2011). Performance analysis of beamforming algorithms. *International Journal of Electronics and Communication Technology* 2 (1).

65 Lin, M., Ouyang, J., and Zhu, W. (2016). Joint beamforming and power control for device-to-device communications underlaying cellular networks. *IEEE Journal on Selected Areas in Communications* 34 (1): 138–150.

66 Xu, C., Song, L., Han, Z. et al. (2012). Interference-aware resource allocation for device-to-device communications as an underlay using sequential second price auction. *Proceedings of the IEEE International Conference on Communications (ICC).*

67 Salameh, H.B. and Hailat, T. (2015). Iterative beamforming algorithm for improved throughput in multi-cell multi-antenna wireless systems. *IET Communications* 9 (13): 1619–1626.

3

Rethinking D2D Interference: Beyond the Past

Mohammad Asad Rehman Chaudhry and Zakia Asad

Soptimizer, Canada

The notion of interference, an undesired signal coupled with a desired signal, can be traced back to the study of sign processes, for example Ferdinand de Saussure's study of signs. The advent of wireless communication brought the study of interference into the mainstream. Traditionally, the focus has been to design the communication systems that do not cause interference which resulted in the development of a wide spectrum of orthogonalization techniques, for example time and frequency domain orthogonalization. Within the last decade and so, theoreticians and practitioners have developed novel methods to exploit interference, instead, and have shown the advantage of such methods over the traditional approaches. These methods utilize a large spectrum of tools spanning from communication theory to theoretical computer science. Our objective is to provide the reader with a thorough and insightful overview of this development. Consequently, we define the Interference Manipulation problem which overlaps with many of the new approaches for interference exploitation relevant to device-to-device (D2D) scenarios. We use tools from different domains to dissect Interference Manipulation, draw its connections with other well-known problems, and deduce solutions.

3.1 Interference Manipulation

IDC forecasts that there would be more than 41.6 billion connected devices, generating 79.4 zettabytes of data in 2025 [1]. On the one hand, these devices are expected to bring revolutionary comfort to our lives, but on the other hand there would be enormous challenges in terms of storage, transmission, privacy, security, authenticity, and meaning of the data generated by these devices. Interference mitigation lies at the heart of all these challenges [2–4]. Specifically, in contrast to traditional approaches advocating interference avoidance as an effective way, this chapter discusses the concept of Interference Manipulation.

Interference Mitigation in Device-to-Device Communications, First Edition.
Edited by Masood Ur Rehman, Ghazanfar Ali Safdar, and Mohammad Asad Rehman Chaudhry.

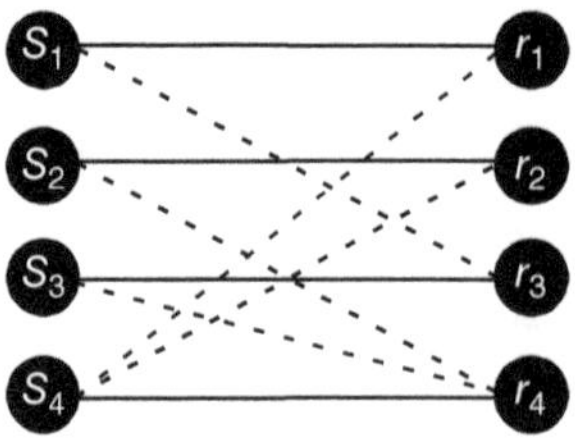

Figure 3.1 A multi-source multi-receiver wireless interference network.

3.1.1 Example

We start by a presenting a simple yet comprehensive example of Figure 3.1. This example introduces *Interference Manipulation* problem, where the objective is to maximize the average achievable data rate – minimize number of channel uses – in an interference network. Specifically, the multi-source multi-receiver wireless network of Figure 3.1 consists of four sources $s_1,\ s_2,\ s_3,\ s_4$, four packets (symbols) $p_1,\ p_2,\ p_3,\ p_4$, and four receivers $r_1,\ r_2,\ r_3,\ r_4$.

Each source s_i transmits packet p_i destined for the receiver r_i, where $i = 1, 2, 3, 4$. For instance, s_1 transmits packet p_1 for the receiver r_1. Any packet p_j received by the destination r_i with $j \neq i$ is interference. For instance, p_4 is an inference packet for r_1. The end-to-end transfer functions for this network can be represented as below for a quasi-static channel use.

$$\begin{aligned}
y_1 &= h_{11}p_1 + h_{14}p_4 \\
y_2 &= h_{22}p_2 + h_{24}p_4 \\
y_3 &= h_{31}p_1 + h_{33}p_3 \\
y_4 &= h_{42}p_2 + h_{43}p_3 + h_{44}p_4
\end{aligned} \tag{3.1}$$

where y_i is the transmission received by receiver r_i, h_{ij} is the nonzero channel coefficient between receiver r_i, and source s_j. Moreover, all the operations are performed over a finite Galois field. The transmitter power constraints are chosen such that each transmitter is able to maintain a desired signal-to-noise-ratio, and *successfully* transmit one packet per channel use in the absence of interference. A source's transmission is said to be successful if intended the receiver can identify the transmitted packet. We want to point out that such a scenario and Interference Manipulation problem overlap with and capture many of the well-known techniques in interference exploitation [5–13].

It is easier to see that in the absence of any interference, each source can successfully transmit a packet per channel use, and there would be a total of four successfully transmissions per channel use giving average achievable data rate of 1 packets per receiver.

On the other, in the presence of interference – as specified by Eq. (3.1) – using traditional interference avoidance, only two sources can make successful transmissions per channel use. Consequently, at least three channel uses are needed for successful transmission of all the four packets to their intended receivers, i.e. an average achievable data rate of $\frac{1}{3}$ per receiver. For instance in a time-division-based orthogonalization scheme sources, s_1 and s_2 transmit during the first time use as their transmissions do not interfere with each other. In the second channel use, s_3 transmits alone. In the third channel use, s_4 transmits alone. The following equations show the packets received at each receiver during each channel use. Noting that channel coefficients h_{ij} are constant, each receiver can identify the received packet.

First channel use:

$$\begin{aligned} y_1 &= h_{11}p_1 \\ y_2 &= h_{22}p_2 \end{aligned} \tag{3.2}$$

Second channel use:

$$y_3 = h_{33}p_3 \tag{3.3}$$

Third channel use:

$$y_4 = h_{44}p_4 \tag{3.4}$$

Next, we ask the question can we do better? The answer is yes. Specifically, instead of avoiding the *mixing* nature of an interference channel, we can use it smartly to do better. As shown in Table 3.1, the sources s_1, s_2, and s_3 transmit during the first channel use, and the sources s_1 and s_4 transmit during the second channel use. This results in an average achievable data rate of $\frac{1}{2}$ per receiver, which is an improvement of 50% compared to the scheme focused on total interference avoidance.

The following equations show the packets received at each receiver during each channel use.

First channel use:

$$\begin{aligned} y_1 &= h_{11}p_1 \\ y_2 &= h_{22}p_2 \\ y_3 &= h_{31}p_1 + h_{33}p_3 \\ y_4 &= h_{42}p_2 + h_{43}p_3 \end{aligned} \tag{3.5}$$

Second channel use:

$$\begin{aligned} y_1 &= h_{11}p_1 + h_{14}p_4 \\ y_2 &= h_{24}p_4 \\ y_3 &= h_{31}p_1 \\ y_4 &= h_{44}p_4 \end{aligned} \tag{3.6}$$

Table 3.1 Interference manipulation.

Source	**First channel use**	**Second channel use**
s_1	Transmit	Transmit
s_2	Transmit	
s_3	Transmit	
s_4		Transmit

It is easier to see that the sources s_1 and s_2 are able to make successful transmissions to their destined receivers in the first channel use. Similarly, the source s_4 is able to make a successful transmissions to its destined receiver in the second channel use. r_3, the intended receiver for source s_3, can make use of the two mixed signals (transmissions containing interference) it received during the first and second channel use to recover its intended packet. Specifically, r_3 receives $h_{31}p_1 + h_{33}p_3$ during the first channel use, and $h_{31}p_1$ during the second channel use. Combining these two enables r_3 to recover its intended packet p_3 as shown in Eq. (3.7), where $\ominus$ represents additive inverse.

$$[h_{31}p_1 + h_{33}p_3] \ominus [h_{31}p_1] \tag{3.7}$$

3.2 Formulation of Interference Manipulation Problem

An instance of Interference Manipulation problem is defined by a set of ℓ sources $s_1, \dots, s_\ell$ and ℓ packets $p_1, \dots, p_\ell$ that are to be delivered to m receivers $r_1, \dots, r_m$. Let the set of sources that transmit during channel use ω is represented by $s_\phi^\omega, \dots, s_\theta^\omega$. Similarly, the receivers that receive some transmission(s) during channel use ω are represented by $r_\alpha^\omega, \dots, r_\rho^\omega$. Note, during channel use ω a receiver $r_j \in r_\alpha^\omega, \dots, r_\rho^\omega$ receives a transmission – y_j^ω – from a subset of sources belonging to $s_\phi^\omega, \dots, s_\theta^\omega$ if either r_i is intended receiver for those or they are a source of interference for them as shown in Eq. (3.8). Without the loss of generality, we assume that a source transmits one packet, if it is transmitting more than one packets we can replace it with several sources each transmitting just one packet.

$$y_j^\omega = \sum_{n=\phi}^{\theta} g_{nj}^\omega \, p_n \tag{3.8}$$

where $g_{nj}^\omega \in GF(q)$ (Galois Field of size q) and is the coefficient of packet p_n received at receiver r_j during transmission ω. For example in Eq. (3.5) of Example 3.1.1, $g_{11}^\omega = h_{11}$.

A receiver r_j is able to recover the packet p_j destined for it, if p_j lies in the subspace spanned by the y_j^ω, where $\omega = 1, \dots, k$ as given by Eq. (3.9).

$$p_j = \sum_{\omega=1}^{k} f_j^\omega \, y_j^\omega \tag{3.9}$$

where $f_j^\omega \in GF(q)$. For example in Eq. (3.7) of Example 3.1.1, r_3 recovers p_3 by performing the operation given by Eq. (3.10), where $f_3^1 = f_3^2 = 1$.

$$f_3^1[h_{31}p_1 + h_{33}p_3] \ominus f_3^2[h_{31}p_1] \tag{3.10}$$

A successful transmission scheme ensures that an equivalent of Eq. (3.9) holds true for all the receivers.

In short, a transmission scheme is optimal if it enables all the sources to successfully transmit to their destined receivers in minimum number of channel uses, say k, as shown in Eq. (3.11).

$$\begin{aligned} &\forall\, p_j : j = 1, \ldots, m,\ p_j = \sum_{\omega=1}^{k} f_j^{\omega}\, y_j^{\omega} \\ &\forall\, k' < k,\ \exists\, p_i \in \{p_1,\ \ldots,\ p_m\}, \ni\ p_i \neq \sum_{\omega=1}^{k'} f_i^{\omega}\, y_i^{\omega} \end{aligned} \tag{3.11}$$

3.3 Matrix Rank Minimization: A Way to Manipulate Interference

We explore the relationship of Interference Manipulation problem with minimum rank of a matrix. We start by defining Matrix Rank Minimization problem.

Definition 3.1 ***(Rank Minimization Problem)*** For a given matrix M of size $i \times \ell$ with all the entries from $GF(q)$, there is a set of unfilled entries x_{jk} (where $j \in 1, 2, \ldots, i$, and $k \in 1, 2, \ldots, \ell$). The objective is to find the values of x_{jk} such that the rank of matrix M is minimized. We represent rank of matrix M by $|M|$.

For example consider the matrix A of size 2×2 as shown in Eq. (3.12), with all the entries from $GF(3)$. The entry at position (2,1) marked as x is missing.

$$A = \begin{pmatrix} 0 & 1 \\ x & 2 \end{pmatrix} \tag{3.12}$$

First note, any value of $x \neq 0$ will result in the second row being linearly independent of the first row, resulting in $|A| = 2$. On the other hand, a choice of $x = 0$ will make the second row linearly dependent on the first row resulting in $|A| = 1$. Hence, in this example, the optimal choice for minimizing rank of A is $x = 0$.

Next, we focus on example of Section 3.1.1. We noticed that Interference Manipulation increased achievable data rate per receiver by 50% compared to the scheme focused on total interference avoidance. The Interference Manipulation scheme made sources s_1, s_2, and s_3 transmit during the first channel use, and the sources s_1 and s_4 transmit during the second channel use. But we did not show how did we find this specific scheme. In fact, we could have used Matrix Rank Minimization to come up with this scheme as described in Section 3.3.1.

3.3.1 Reduction of Interference Manipulation to Matrix Rank Minimization

Specifically, for a given instance of Interference Manipulation problem over a finite field of size q $GF(q)$, we define matrix M – an instance of the Matrix Rank Minimization problem as below:

- For each receiver r_i there is a row i in the matrix M;
- For each source s_j there is a column in the matrix M;
- An entry $m_{i,j} = 1$, if receiver r_i is the intended destination of source s_j;

- An entry $m_{k,\ell} = 0$, if receiver r_k receivers interference from source s_ℓ;
- All the other entries are missing and can take any value from $GF(q)$.

The objective is to choose the values of the missing entries such that the rank of the matrix M is minimized. We can this new minimum rank matrix M′.

For the instance of Interference Manipulation problem given in Section 3.1.1, the corresponding instance of Matrix Rank Minimization problem given by matrix K is given below in Eq. (3.13), where each x_{ij} represents a missing value from $GF(2)$.

$$K = \begin{pmatrix} 1 & x_{12} & x_{13} & 0 \\ x_{21} & 1 & x_{23} & 0 \\ 0 & x_{32} & 1 & x_{34} \\ x_{41} & 0 & 0 & 1 \end{pmatrix} \begin{matrix} r_1 \\ r_2 \\ r_3 \\ r_4 \end{matrix} \quad \text{(columns } s_1, s_2, s_3, s_4\text{)} \tag{3.13}$$

It can be observed that setting $x_{12} = x_{13} = x_{21} = x_{23} = x_{41} = x_{32} = x_{34} = 1$ minimizes the rank of matrix K to 2, as shown in Eq. (3.14). Specifically, it is easy to see that first and second rows are linearly dependent on each other. Moreover, the third row can is linearly dependent on first and fourth row and can be obtained by addition of first and fourth row over $GF(2)$ – bitwise Exclusive OR – i.e. $[1110] + [0111] = [1001]$.

$$K = \begin{pmatrix} 1 & \mathbf{1} & \mathbf{1} & 0 \\ \mathbf{1} & 1 & \mathbf{1} & 0 \\ 0 & \mathbf{1} & 1 & \mathbf{1} \\ \mathbf{1} & 0 & 0 & 1 \end{pmatrix} \begin{matrix} r_1 \\ r_2 \\ r_3 \\ r_4 \end{matrix} \quad \text{(columns } s_1, s_2, s_3, s_4\text{)} \tag{3.14}$$

Next, we move with removing dependent rows from the matrix K, getting matrix K' as shown in Eq. (3.15).

$$K' = \begin{pmatrix} 1 & \mathbf{1} & \mathbf{1} & 0 \\ \mathbf{1} & 0 & 0 & 1 \end{pmatrix} \quad \text{(columns } s_1, s_2, s_3, s_4\text{)} \tag{3.15}$$

3.3.2 Minimum Rank Matrix to Transmission Scheme

This section describes the procedure of finding a transmission scheme that optimizes the average achievable data rate per receiver. Given the minimum rank matrix M' – with rank $|M'|$ – from Section 3.3.1, we construct the transmission scheme as below.

- Each row i in M' represents a channel uses, where $i = 1, 2, \ldots, |M'|$;
- Source s_j transmits during channel use i if the matrix element $m'_{ij} \neq 0$.

To explain this transformation, we use the instance of Interference Manipulation problem given in Section 3.1.1, and minimum rank matrix K' given by Eq. (3.15). Each row of K' corresponds to a channel use as shown by (3.16), i.e. in a total of two channel uses each source can successfully transmit to its intended receiver – an average achievable data rate of $\frac{1}{2}$ per receiver.

$$\begin{matrix} s_1 & s_2 & s_3 & s_4 \\ \end{matrix}$$
$$\begin{pmatrix} 1 & \mathbf{1} & \mathbf{1} & 0 \\ \mathbf{1} & 0 & 0 & 1 \end{pmatrix} \begin{matrix} \text{First channel use} \\ \text{Second channel use} \end{matrix} \tag{3.16}$$

Specifically, sources s_1, s_2, and s_3 transmit during the first channel use, and sources s_1 and s_4 transmit during the second channel use as shown by (3.17).

$$\begin{pmatrix} s_1 & s_2 & s_3 & 0 \\ s_1 & 0 & 0 & s_4 \end{pmatrix} \begin{matrix} \text{First channel use} \\ \text{Second channel use} \end{matrix} \tag{3.17}$$

Moreover, the packets transmitted during each channel use are given by 3.18.

$$\begin{pmatrix} P_1 & P_2 & P_3 & 0 \\ P_1 & 0 & 0 & P_4 \end{pmatrix} \begin{matrix} \text{First channel use} \\ \text{Second channel use} \end{matrix} \tag{3.18}$$

3.3.3 Does the Field Size Matter?

In this section, we investigate the relationship between the size of Galois field and the number of channel uses for Interference Manipulation problem [14]. We show that number of channel uses depend on the field size.

To prove our point, we focus on the example discussed in Sections 3.3.1–3.3.2. We have already shown that the matrix K, Eq. (3.13), has a minimum rank of 2 – if all operations are performed over $GF(2)$ – and accordingly an average achievable data rate of $\frac{1}{2}$ per receiver. On the other hand, if all the operations are performed over $GF(3)$, then the same matrix K has a minimum rank of 3 and accordingly an average achievable data rate of $\frac{1}{3}$ per receiver. That is three channel uses are required, instead of two, when all the operations are performed over $GF(3)$.

3.4 Interference Manipulation: A Boolean Satisfiability Approach

In Section 3.3, we have established the relationship between Interference Manipulation and Matrix Rank Minimization. Matrix Rank Minimization is one of the toughest problems in computational complexity with a solution complexity not feasible for practical scenarios. Specifically, Matrix Rank Minimization is not only nondeterministic polynomial (NP)-hard but NP-hard to even approximate [15]. Consequently, we transform Interference Manipulation problem to Boolean Satisfiability Problem. Boolean Satisfiability problem is an NP-hard problem as well [16], but fortunately some practically useful programs, called SAT-Solvers, have been developed that can solve it for many practical scenarios with reasonable computational complexity.

We begin by an examination of Eqs. (3.8)-(3.11). We then express Eq. (3.8) in its equivalent form given by Eq. (3.19), where $g^{\omega}_{nj} = 0$ corresponds to a packet transmitted by s_i if $s_i \notin s^{\omega}_{\phi}, \dots, s^{\omega}_{\theta}$.

$$y^{\omega}_j = \sum_{n=1}^{\ell} g^{\omega}_{nj}\, p_n \tag{3.19}$$

We further note that Eqs. (3.9) and (3.19) result in the following constraints. Specifically, for each receiver r_j with intended packet p_j Constraint (3.20) holds.

$$\sum_{\omega=1}^{k} g_{nj}^{\omega} f_j^{\omega} \equiv 1 \tag{3.20}$$

Moreover, for each receiver r_j with a non-intended packet – interference – p_j Constraint (3.21) holds.

$$\sum_{\omega=1}^{k} g_{nj}^{\omega} f_j^{\omega} \equiv 0 \tag{3.21}$$

Constraints (3.20) and (3.21) can be transformed into a Boolean form by substituting the summation and multiplication over $GF(2)$ by AND ($\wedge$)and Exclusive OR ($\oplus$) operations, respectively, over the boolean variables. Equations (3.22) and (3.23) are the Boolean transformations of (3.20) and (3.21), respectively.

$$\oplus_{\omega=1}^{k} g_{nj}^{\omega} \wedge f_j^{\omega} \equiv 1 \tag{3.22}$$

$$\oplus_{\omega=1}^{k} g_{nj}^{\omega} \wedge f_j^{\omega} \equiv 0 \tag{3.23}$$

Equations (3.22) and (3.23) can be transformed into the conjunctive normal form (CNF) using the Tustin transformation [17] and then can efficiently be solved for several practical instances by the state-of-the-art SAT-Solvers [18–21].

3.5 Interference Manipulation: Index Coding Perspective

This section digs deep into Interference Manipulation problem by establishing its relationship with Index Coding problem – a fundamental problem in Information Theory [22, 23]. We start by defining an instance of Index Coding problem [7, 12].

Definition 3.2 ***(Index Coding)*** An instance of index coding problem is defined by a relay R which contains a set of ℓ packets $X = \{x_1, \dots, x_\ell\}$ that are to be delivered to a set of m clients $C = \{c_1, \dots, c_m\}$ over a *broadcast channel.* Each client c_i has access to some *side information* $HIC_i \subseteq X$ and requires a set of packets $WIC_i \subseteq X$ from the relay R. The relay R can transmit packets in X or their combinations over $GF(q)$. The objective is to find a scheme that requires the minimum number of transmissions from the relay R and satisfies the requests of all the clients. Let $OPT(IC)$ denote the optimal solution of Index Coding problem, and $|OPT(IC)|$ denote the minimum number of transmissions from the relay.

Given an instance of Interference Manipulation problem, we define an instance of Index Coding problem as follows.

- $C = \{r_1, \dots, r_m\}$, where $r_i = c_i \ \forall i$;
- $X = \{p_1, \dots, p_\ell\}$, where $x_i = p_i \ \forall i$;
- $R = \bigcup_{i=1}^{\ell} s_i$;

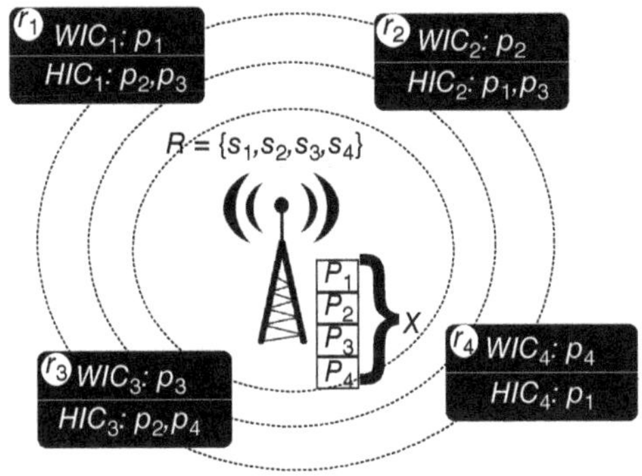

Figure 3.2 Index coding's instance corresponding to Interference Manipulation's instance depicted in Figure 3.1.

- $WIC_i = \bigcup p_i$, where p_i is destined for r_i;
- $HIC_i = \bigcup p_j$, where p_j is neither destined for r_i nor causes an interference to it.

For the instance of Interference Manipulation problem given in Section 3.1.1, the corresponding instance of Index Coding problem is shown in Figure 3.2 and also given by Eq. (3.24).

$$
\begin{aligned}
C &= \{r_1, r_2, r_3, r_4\} \\
X &= \{p_1, p_2, p_3, p_4\} \\
R &= \{s_1, s_2, s_3, s_4\} \\
WIC_1 &= \{p_1\} \\
HIC_1 &= \{p_2, p_3\} \\
WIC_2 &= \{p_2\} \\
HIC_2 &= \{p_1, p_3\} \\
WIC_3 &= \{p_3\} \\
HIC_3 &= \{p_2, p_4\} \\
WIC_4 &= \{p_4\} \\
HIC_4 &= \{p_1\}
\end{aligned} \tag{3.24}
$$

3.5.1 Interference Manipulation Is NP-hard

In this section, we show that both Interference Manipulation problem and Index Coding problem are equivalent to each other. Recall, for Interference Manipulation problem, a transmission scheme is optimal if it enables all the sources to successfully transmit to their destined receivers in minimum number of channel uses as shown in Eq. (3.11). We call this scheme *OPT*(*IM*), and |*OPT*(*IM*)| is the total number of channel uses, represent by k in Section 3.1.1. We show that |*OPT*(*IM*)| = |*OPT*(*IC*)|. The proof follows from the observation that there exists a one-to-one mapping between elements of Interference Manipulation problem and Index Coding problem. Specifically, we note that each transmission from relay R in Index Coding represents a channel use for Interference Manipulation problem. Moreover, we note that non-interfering transmissions in Interference Management are treated as side information in Index Coding which can be been seen from Eq. (3.20). Furthermore, interfering transmissions in Interference Management are treated as non-desired packets in Index Coding which is obvious from Eq. (3.21). Additionally, the optimal solution to Index Coding $|OPT(IC)|$ results in minimizing the number of channel uses in Interference Manipulation $|OPT(IM)|$.

The equivalence between Interference Manipulation problem and Index Coding problem also provides us with the computational complexity of finding the optimal solution to Interference Manipulation problem. Index coding problem has been shown to be NP-hard and NP-hard to approximate [7], and we have shown Interference Manipulation problem to be equivalent to it. Hence, *Interference Manipulation problem is NP-hard, and it is NP-hard to find a polynomial time constant factor approximation for it as well.*

3.5.2 An Efficient Solution for Interference Manipulation

In this section, we present a time-efficient – polynomial time – solution for Interference Manipulation problem. Such a solution could be viable for large instances encountered in D2D communication.

We transform an instance of Interference Manipulation to an equivalent instance of Index Coding problem. Then we solve the instance of Index Coding problem by using the *Cycle Packing*-based algorithm [24]. Finally, we deduce a transmission schedule for different sources based on the solution of Index Coding problem.

We start by introducing a notion of the *Dependency Graph*, which captures the relationship between receivers, their desired packets, and interference. The dependency graph is defined for the multiple unicast case, in which each packet is requested by a single client.

Definition 3.3 ***(Dependency Graph)*** Given an instance of the Index coding problem, we define a graph $G(V, E)$ as follows:

- For each receiver $r_i \in C$, there is a corresponding vertex v_{c_i} in V
- There is a directed edge from v_{c_i} to v_{c_j} if and only if it holds that $WIC_i \in HIC_j$.

Figure 3.3a depicts the dependency graph that corresponds to the instance of the *IC* problem depicted in Figure 3.2. Figure 3.4 shows the overall flow of this solution.

The main idea of Algorithm *IM* is to find a disjoint cycles in the dependency graph. It is interesting to note that all the receivers connected by a directed cycle of length φ' in dependency graph can recover their desired packets in $\varphi' - 1$ channel uses [7]. To see this, consider dependency graph depicted in Figure 3.3b. This graph only contains one cycle consisting of three receivers r_1, r_2, and r_3. All these three receivers can recover their required packers p_1, p_2, and p_3, respectively, in two channels uses. Such that sources s_1 and s_2 transmit in the first channel use, and sources s_2 and s_3 transmit during the second channel use. Each of r_1, r_2, and r_3 can recover the packet destined to it by combining the transmissions it received during these two channel uses by deploying the approach discussed in earlier parts of this chapter, see

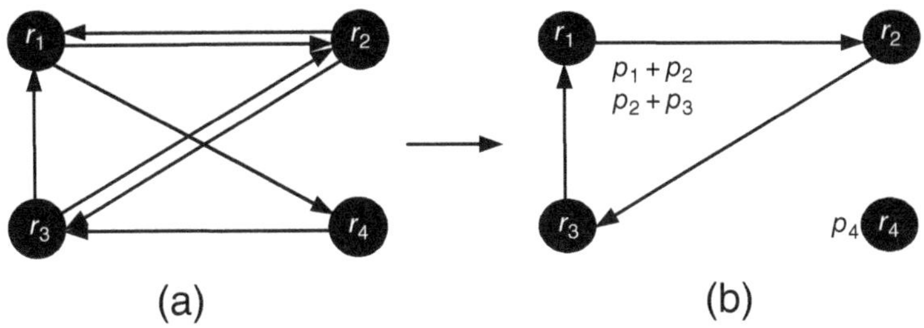

Figure 3.3 (a) Dependency graph for the instance of the Index Coding problem depicted in Figure 3.2. (b) Disjoint cycle packing on the dependency graph and corresponding set of transmissions.

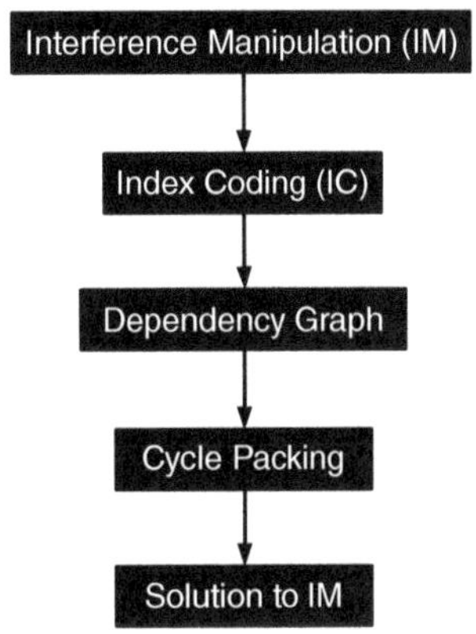

Figure 3.4 An efficient solution for Interference Manipulation.

Algorithm *IM*:
1 From the given instance of the *Index Coding* problem, construct the *Dependency Graph G(V, E)*;
2 $\varphi = \emptyset$
3 **While** there exists a directed cycle in $G(V, E)$ **do**:
4 Find a cycle φ' of minimum length
5 Add φ' to φ
6 Delete all the edges of *varphi'* from $G(V, E)$
7 **For** each cycle $\varphi' \in \varphi$ **do**:
8 Transmit a set of $|\varphi'| - 1$ transmissions that satisfy all r_i in φ'
9 For each $v_i \in G(V, E)$ not included in any cycle φ', transmit the packet required by the corresponding client r_i

Figure 3.5 Algorithm *IM*.

for instance Eq. (3.9). Furthermore, for receiver r_4 which is not a part of any cycle, source s_4 transmits during the third channel use. In short, all the four receivers can receive the packets destined toward them in three channel uses, giving an average achievable rate of $\frac{1}{3}$ per receiver. This average achievable rate is less than optimal achievable data rate given by the optimal solutions presented in Sections 3.3 and 3.4. On the other hand, compared to the solutions provided in Sections 3.3 and 3.4 that are NP-hard, the cycle packing based solution is extremely efficient in terms of computational complexity – polynomial time and is feasible for instances where a large number of devices are in constant communication over an interference network.

3.6 Summary

Interference is one of the major issues hampering the D2D communication. In this chapter, we have investigated possibilities and methods available to manipulate the interference and minimize its impact. We formulated the Interference Manipulation problem in D2D communication, which overlaps with many of the new approaches for interference exploitation relevant to D2D scenarios. Methods and techniques including matrix rank minimization, Boolean satisfiability, and index coding have been reviewed for their usability in Interference Manipulation along with their connections with other well-known problems.

References

1 Shirer, M. and MacGillivray, C. (2019). The growth in connected IoT devices is expected to generate 79.4 ZB of data in 2025, according to a new IDC forecast.

2 Kwan, R. and Leung, C. (2010). A survey of scheduling and interference mitigation in LTE. *Journal of Electrical and Computer Engineering* 2010, Article ID 273486, 10 p.

3 Gummadi, R., Patra, R.K., Balakrishnan, H., and Brewer, E.A. (2008). Interference avoidance and control. *HotNets*, pp. 13–18.

4 Dai, X., Chen, S., Sun, S. et al. (2014). Successive interference cancelation amenable multiple access (SAMA) for future wireless communications. *2014 IEEE International Conference on Communication Systems*, pp. 222–226. IEEE.

5 Jafar, S.A. (2014). Topological interference management through index coding. *IEEE Transactions on Information Theory* 60 (1): 529–568.

6 Sun, H., Geng, C., and Jafar, S.A. (2013). Topological interference management with alternating connectivity. *2013 IEEE International Symposium on Information Theory*, pp. 399–403. IEEE.

7 Chaudhry, M.A.R. (2011). Network coding in distributed, dynamic, and wireless environments: algorithms and applications. PhD thesis. Texas A&M.

8 Asad, Z., Chaudhry, M.A.R., and Malone, D. (2016). Greener data exchange in the cloud: a coding-based optimization for big data processing. *IEEE Journal on Selected Areas in Communications* 34 (5): 1360–1377.

9 Lee, N. and Heath, R.W. Jr. (2016). Advanced interference management technique: potentials and limitations. *IEEE Wireless Communications* 23 (3): 30–38.

10 Ahlswede, R., Cai, N., Li, S.-Y.R., and Yeung, R.W. (2000). Network information flow. *IEEE Transactions on Information Theory* 46 (4): 1204–1216.

11 Chaudhry, M.A.R., Asad, Z., Sprintson, A., and Hu, J. (2011). Efficient congestion mitigation using congestion-aware Steiner trees and network coding topologies. *VLSI Design*.

12 Chaudhry, M.A.R. and Sprintson, A. (2008). Efficient algorithms for index coding. *IEEE Conference on Computer Communications (INFOCOM) Workshops*, pp. 1–4.

13 Asad, Z. (2017). Greener big data: optimizing data exchange and power procurement for data centers. PhD thesis. University of Toronto.

14 El Rouayheb, S.Y., Chaudhry, M.A.R., and Sprintson, A. (2007). On the minimum number of transmissions in single-hop wireless coding networks. *2007 IEEE Information Theory Workshop*, pp. 120–125. IEEE.

15 Candès, E.J. and Recht, B. (2009). Exact matrix completion via convex optimization. *Foundations of Computational Mathematics* 9 (6): 717.

16 Cook, S.A. (1971). The complexity of theorem-proving procedures. Proceedings of the 3rd Annual ACM Symposium on Theory of Computing, pp. 151–158.

17 Tseitin, G.S. (1983). On the complexity of derivation in propositional calculus. In: J.H. Siekmann, G. Wrightson (eds) *Automation of Reasoning. Symbolic Computation (Artificial Intelligence)*, pp. 466–483. Springer: Springer, Berlin, Heidelberg.

18 Fazekas, K. (2020). On SAT-based solution methods for computational problems. PhD thesis. Informatik, Johannes Kepler University Linz.

19 Biere, A., Fazekas, K., Fleury, M., and Heisinger, M. (2020). CaDiCaL, Kissat, Paracooba, Plingeling and Treengeling entering the SAT Competition 2020, pp. 51–53.

20 Biere, A. (2019). CaDiCaL at the SAT Race 2019. *Proceedings of SAT Race 2019 – Solver and Benchmark Descriptions*, pp. 8–9.

21 Hossen, M.S. and Polash, M.M.A. (2019). Implementing an efficient SAT solver for structured instances. *2019 Joint 8th International Conference on Informatics, Electronics & Vision (ICIEV) and 2019 3rd International Conference on Imaging, Vision & Pattern Recognition (icIVPR)*, pp. 238–242. IEEE.

22 Birk, Y. and Kol, T. (1998). Informed-Source Coding-on-Demand (ISCOD) over broadcast channels. *Proceedings of INFOCOM'98*.

23 Bar-Yossef, Z., Birk, Y., Jayram, T.S., and Kol, T. (2006). Index coding with side information. *Proceedings of 47th Annual IEEE Symposium on Foundations of Computer Science(FOCS)*, pp. 197–206.

24 Chaudhry, M.A.R., Asad, Z., Sprintson, A., and Langberg, M. (2011). On the complementary index coding problem. *2011 IEEE International Symposium on Information Theory Proceedings*, pp. 244–248. IEEE.

4

User Pairing Scheme for Efficient D2D Content Delivery in Cellular Networks

Yanli Xu

College of Information Engineering, Shanghai Maritime University, Shanghai, China

Considering that future cellular networks are more content-centric and content sharing among user equipments (UEs) is more universal, device-to-device (D2D) communication techniques become attractive for their applications to content delivery due to the characteristic of proximal transmissions without the needs of high-rate backhaul. How to efficiently pair a content transmitter with potential receivers is a primary problem for D2D content delivery. This problem becomes more important when the transmitter explores a multicast mode, where the user paring (a transmitter with multiple receivers) will affect both the interference on other transmitters and the efficiency of content delivery. In this chapter, we will discuss the user paring range (multicast range) for D2D multicast content delivery. Then we analyze the optimal paring range. Based on the analyses, we further discuss the optimal multicast strategy and caching strategy to maximize the benefit of D2D user paring. Lastly, we give an evaluation metric to determine when to use the user paring scheme. From the illustration of this chapter, we can see that the D2D paring and content delivery can be designed as an energy-efficient one which minimizes the transmitter power for a targeted performance. As a result, the interference of content delivery can be minimized.

4.1 D2D Content Delivery

Recently, data traffic grows rapidly due to the development of mobile devices and wireless communication technologies. Unlike the previous growth of mobile traffic, which mainly depended on the demands of connection-centric communications, such as phone calls and text messages, the recent and future growth would come from the explosion of content-centric communications such as video streaming and content sharing. However, content delivery aggravates the burden of the core network and is not energy-efficient due to frequent content retrieval from base stations (BSs).

Interference Mitigation in Device-to-Device Communications, First Edition.
Edited by Masood Ur Rehman, Ghazanfar Ali Safdar, and Mohammad Asad Rehman Chaudhry.

Toward green fifth-generation (5G) cellular networks, some researchers have studied the utilization of small-cell BSs with large storage units (such as hard disks and solid-state drives) for content delivery to reduce the resource consumption by bringing content closer to users [1–4]. For example, system performance metrics such as outage probability and content delivery rate were analyzed in [1]. Caching strategies about how to store data at small-cell BSs were studied in [2–4]. However, such small-cell architecture may cause bottlenecks of content delivery due to the requirement of high-rate backhaul links as indicated in [5]. In D2D communications, data are directly transmitted between UEs without the traverse of BSs and the transmission distance is usually short [6, 7], which will improve the resource efficiency. In addition, there is a large amount of content reuse among UEs, especially for popular contents. Thereby, D2D communications for content delivery have attracted great attention recently [8–10].

In [8], D2D pairs with common interests construct a cluster for D2D content sharing, where the number of requests from D2D UEs that could be served in a cluster, and the optimized global density of served requests were analyzed. The work [9] analyzed the optimal cluster size for random and fixed caching via assuming that users cache popular video files and send them after receiving requests from other users in the same cluster. A tractable closed-form equation to find when redundant caching should be used in order to minimize the expected energy consumption was derived in [10]. However, these research works were based on a simplistic system model with the assumptions which did not accurately consider the effects of D2D interference or fading. Thus, some works analyzed the D2D content delivery problem based on stochastic geometry theory [11] with the Poisson Point Process (PPP) UE distribution [12–14]. For example, caching policy was studied for D2D communications in millimeter-wave (mmWave) networks, and the offloading gain of the proposed policy was analytically derived based on stochastic geometry theory in [12]. The work [13] optimized the D2D caching to maximize the total success probability of content delivery. The work [14] analyzed sharing content failure rate in a certain range and proposed an algorithm to determine the probability that content is cached in a server to minimize the failure probability. Particularly in [15], based on social relationships among mobile users and a heterogeneous but overlapping interests of users model, a graph-based social-aware algorithm was proposed, in which cellular links and D2D links were established according to social ties and social contributions of users for efficient multi-file delivery. Through integrating network coding with D2D communication, Pyattaev et al. [16] demonstrated the benefits of network coding for D2D content delivery, and that the performance can be improved even by simple network coding.

Note that the architecture of D2D content delivery was based on request-and-response pattern in most existing proposals. As advocated in [17], the multicast gain can play an important role in proactive caching for the 5G cellular networks. Realizing the advantage of multicast, UE multicast was used for content recovery of UEs in the same multicast group as an assistant method of cellular multicast in [18–20]. For content delivery based on D2D multicast, a model based on PPP was proposed and the system performance was analyzed in [21]. Considering that proximal UEs have clustering distribution and there may be multiple potential transmitters in a cluster, a multicast model based on Poisson cluster process was investigated for D2D cache and delivery in [22]. To distinguish appropriate conditions of employing unicast and multicast, performance metrics under different transmission environment parameters are investigated in [23]. In addition to the content delivery pattern, the energy efficiency of D2D content delivery needs to be considered for the sake of green 5G [24]. Toward energy-efficient content delivery, centralized and distributed Medium Access Control (MAC) strategies are studied in [25, 26] to achieve balance between conserving energy and content delivery. In [27], social awareness was exploited in order to improve the energy efficiency of D2D cooperative communication. Except above methods, user paring is an important way for efficient D2D content delivery. It improves delivery efficiency by pairing D2D transmitters with appropriate number of receivers. Given paired D2D users, the selection of cached and delivered content can further improve delivery efficiency. These problems will be considered in the following subsections.

4.2 D2D Content Delivery Architecture

4.2.1 Network Model

We consider a cellular network where UEs are randomly located. UEs transmit data under either cellular mode or D2D mode. The transmission mode of a UE is selected by the BS based on a given mode selection criterion. Two UEs are neighbors if the Euclidean distance between them is smaller than a distance threshold and may communicate in the D2D mode. This maximally allowable distance for D2D communication is determined by the transmission power and the used power control scheme of transmitters. There are two ways for the in-band coexistence of different modes, i.e. D2D UEs reusing resource of cellular UEs (underlaying coexistence) and using the orthogonal resource of cellular UEs (overlaying coexistence), respectively. Both of coexistent ways have been widely studied in prior works [28–30]. Compared to orthogonal resource allocation, a resource reuse scheme may improve spectrum efficiency while leads to severer interference and higher implementation complexity. In this chapter, we focus on the efficiency of orthogonal resource allocation case and readers can refer to prior works on the spectrum efficiency of different coexistence ways such as [31–33]. For simultaneous D2D transmitters, uplink resources are spatially reused to save bandwidth since interference among them can be well controlled due to the short transmission range of these communication links. We can use stochastic geometry for the modeling of UE distributions for D2D multicast. Particularly, the positions of multicast D2D transmitters form an independent PPP Ψ_t with density λ_t. For each D2D transmitter, the positions of its intended receivers are independently uniformly scattered around the transmitter in the multicast range which is centered at the transmitter with radius d_0. Thereby, the receiver density of a multicast transmitter is a constant independent to d_0, which is denoted as λ_r here.

4.2.2 Channel Model

We assume that UEs are half-duplex constraint. All UEs are synchronized with BSs. For a communication link $i \to j$, the signal received at the node j can be expressed as:

$$y_j = \sqrt{E_i} h_{ij} x_i + n_0 \tag{4.1}$$

where E_i denotes the transmission power of node i, and n_0 is the thermal noise. h_{ij} denotes the channel coefficient between nodes i and j, which satisfies $\left|h_{ij}\right|^2 = \left|h'_{ij}\right|^2 d_{ij}^{-\alpha}$, h'_{ij} characterizes the rapid fading, d_{ij} is the distance between i and j, and α is the path loss exponent. It is assumed that the channel undergoes the flat Rayleigh fading and h'_{ij} follows zero-mean circular symmetric complex Gaussian distribution $\mathcal{CN}(0, 1/\mu_{ij})$. Signal-interference-ratio (SIR) distribution is independent of μ_{ij} in the interference-limited system, while μ_{ij} will affect the system performance when the noise is considered. Here, we let $\mu_{ij} = 1$ for simplification when noise is ignored.

In this chapter, we randomly take a transmitter i and its corresponding receivers as an example to illustrate the analysis process of the performance of D2D content delivery. Considering that the transmitter is the same for these receivers, with a slight abuse of notation, we omit i when we analyze the multicast performance of a receiver of i in the left part of this chapter, e.g. receiving signal strength, signal to interference and noise ratio (SINR) and the probability that the multicast file meeting the requirement of a receiver. For a receiver j, the received SINR γ_j can be expressed as:

$$\gamma_j = \frac{E_i d_{ij}^{-\alpha} H_{ij}}{I_j + N_0} \tag{4.2}$$

where I_j is the interference at the receiver j and $H_{ij} = \left|h'_{ij}\right|^2$.

Then, a successful transmission of the link $i \to j$ needs to satisfy

$$\gamma_j \geq \gamma_{\text{th}} \tag{4.3}$$

where γ_{th} is the required SINR for an effective communication.

4.2.3 Content Delivery Model

Each UE has a cache with size $\tilde{N}$, which is populated with some video files with the same size. We assume that the probability that a UE caches a file n is $f_c(n)$, and the distribution of cached files is independent identically distributed (iid) at each UE. Not all of these cached contents are required by the UE itself. However, it may provide useful contents to its neighbors by D2D multicast. Similarly, this UE can also benefit from the content cached by other UEs. Hence, this caching policy is mutually beneficial. Assuming that all contents have the same size, UEs disseminate these cached contents to nearby UEs in their transmission resources to bring popular contents closer to users. Here, two ways for content delivery can be utilized according to a predefined mode selection strategy, i.e. D2D delivery and cellular delivery.

The predefined delivery strategy is that UE or BS transmits content following a certain distribution f_s, which is iid at each UE and BS. Based on this distribution, the transmission probability of a content in a cache is determined. For example, multicast transmitters transmit file n with probability $f_s(n)$ in allocated bandwidth during a multicast period. All neighbor UEs of a multicast transmitter will be potential receivers. These potential receivers which have no data to transmit will be scheduled by BS to listen the multicast files of their neighbor transmitters. When a UE is a potential receiver of multiple multicast transmitters, BS randomly selects a multicast transmitter for it. The neighbor relationship among UEs can be obtained by BS during the neighbor discovery period which is an essential procedure before D2D communications standardized by the Third Generation Partnership Project (3GPP) [34].

We use popularity to describe the probability of a file being required by a UE. That is, the required statistics for video files are modeled by a Zipf distribution, which has been shown to fit well with video requirement and widely used for the analysis of D2D caching [35–37]. Hence, the probability $f_r(n)$ that file n is required by UE can be expressed by:

$$f_r(n) = \frac{1}{n^\beta}\left[\sum_{l=1}^{N}\frac{1}{l^\beta}\right]^{-1} \tag{4.4}$$

where β is the Zipf exponent, n is the nth most popular video, and N is the number of files ($N \leq \tilde{N}$).

4.3 D2D Content Delivery Strategies

4.3.1 Pairing Range

We call a UE, which can benefit from the content delivery by receiving its required file, as a benefiting user equipment (BUE). For D2D delivery, the number of BUEs is one of the key evaluation metrices to determine the efficiency of delivery. Since the number of BUEs depends on the paring range of a transmitter, we first give the following definition.

Definition 4.1 A targeted sharing zone (TSZ) is the area in which all UEs can pair with the multicast UE and share with its cached files.

For a transmitter i, its TSZ $\mathcal{T}_i$ can be modeled as a circle with center i and radius d_0. We assume that the path loss is compensated by power control, i.e.

$$E_i = \eta_0 d_0^{k\alpha} \tag{4.5}$$

where k is the path loss compensatory exponent, and η_0 is the targeted receiving power at the receiver.

Since the goal of D2D caching and delivery is to meet file requirement of UEs in the network through proximal transmission, a UE having no data to transmit is scheduled to receive files as long as it locates in the TSZ of a transmitter. Several transmitters may want to pair with a receiver simultaneously, and the receiver randomly selects a transmitter to pair. Since simultaneous transmitters spatially reuse transmission resources, the signals from different transmitters may be interfered with each other at the receiver. For a receiver j and its transmitter i, the SINR of the received signal γ_j at j can be expressed as:

$$\begin{aligned}\gamma_j &= \frac{E_i H_{ij} d_{ij}^{-\alpha}}{I_j + N_0} \\ &= \frac{\eta_0 d_0^{k\alpha} H_{ij} d_{ij}^{-\alpha}}{\sum\limits_{l\in\Psi_t\backslash i} \eta_0 d_0^{k\alpha} H_{lj} d_{lj}^{-\alpha} + N_0} \\ &= \frac{H_{ij} d_{ij}^{-\alpha}}{\sum\limits_{l\in\Psi_t\backslash i} H_{lj} d_{lj}^{-\alpha} + \frac{N_0}{\eta_0 d_0^{k\alpha}}}\end{aligned} \tag{4.6}$$

Denote the file sent by transmitter i as S_i and the file required by receiver j as R_j. When i transmits file n, the conditional probability $M_j(n)$ that n meets the requirement of j is

$$M_j(n) = \Pr\left\{\gamma_j \geq \gamma_{\text{th}}\right\} \Pr\left\{R_j = n | S_j = n\right\} \tag{4.7}$$

where $\Pr\left\{R_j = n | S_j = n\right\}$ is the conditional probability that UE j requires file n when UE i sends this file.

We assume that H_{ij} is exponentially distributed and noise is negligible. The first part in (4.7) can be calculated by (4.8), i.e.

$$\begin{aligned}\Pr\left\{\gamma_j \geq \gamma_{\text{th}}\right\} &= \Pr\left\{H_{ij} \geq \gamma_{\text{th}} d_{ij}^{\alpha} \sum_{l\in\Psi_t} d_{lj}^{-\alpha} H_{lj}\right\} \\ &= \mathrm{E}\left[\exp\left(-\gamma_{\text{th}} d_{ij}^{\alpha} I'_j\right)\right] \\ &= \mathcal{L}_{I'_j}\left(\gamma_{\text{th}} d_{ij}^{\alpha}\right) \\ &= \exp\left(-\pi\lambda_t m\Gamma(m)(1-m)\gamma_{\text{th}}^{m} d_{ij}^{2}\right) \\ &= \exp\left(-\kappa d_{ij}^{2}\right)\end{aligned} \tag{4.8}$$

where $\mathcal{L}_{I'_j}$ is the Laplace transform of $I'_j = \sum_{l\in\Psi_t} d_{lj}^{-\alpha} H_{lj}$, $\kappa = \pi\lambda_t m\Gamma(m)(1-m)\gamma_{\text{th}}^{m}$, $m = \frac{2}{\alpha}$ and $\Gamma(\cdot)$ is the Gamma function. The impact of interference from simultaneous transmitters is implicitly embodied by κ based on stochastic geometry analyses (Appendix A in [38]), which depends on the density of simultaneous transmitters (λ_t). That is, κ increases with interference which is led by the increase in the density of simultaneous transmitters. From (4.8), we see that the average successful transmission is affected by path loss (determined by link distance) and the density of simultaneous transmitters (determined by scheduling).

The second part in Eq. (4.7) can be derived by:

$$\begin{aligned}\Pr\left\{R_j = n | S_j = n\right\} &= \frac{\Pr\left\{R_j = n, S_j = n\right\}}{\Pr\left\{S_j = n\right\}} \\ &= \Pr\left\{R_j = n\right\} \\ &= \frac{1}{n^{\beta}}\left[\sum_{l=1}^{N} \frac{1}{l^{\beta}}\right]^{-1}\end{aligned} \tag{4.9}$$

Including Eqs. (4.8) and (4.9) into Eq. (4.7) yields the conditional probability that file n meets the requirement at j as follows:

$$M_j(n) = \frac{\exp\left(-\kappa d_{ij}^2\right)\frac{1}{n^\beta}}{\sum_{l=1}^{N}\frac{1}{l^\beta}} \tag{4.10}$$

For the receiver j, the probability ω_j that its receiving file meets its requirement when it is in the TSZ of transmitter i can be given by:

$$\omega_j = \sum_{n=1}^{N} \Pr\left\{S_j = n\right\} M_j(n) \tag{4.11}$$

where $\Pr\left\{S_j = n\right\}$ is the probability that transmitter i caches file n and sends it, which can be written as:

$$\Pr\left\{S_j = n\right\} = f_c(n) f_s(n) \tag{4.12}$$

where $f_c(n)$ and $f_s(n)$ are probabilities that the transmitter caches file n and sends it, respectively.

Including Eqs. (4.10) and (4.12) into Eq. (4.11), we have

$$\begin{aligned}\omega_j &= \sum_{n=1}^{N} \Pr\left\{S_j = n\right\} M_j(n)\\ &= \sum_{n=1}^{N} f_c(n) f_s(n)\frac{\exp\left(-\kappa d_{ij}^2\right)\frac{1}{n^\beta}}{\sum_{l=1}^{N}\frac{1}{l^\beta}}\\ &= \frac{\exp\left(-\kappa d_{ij}^2\right)}{\sum_{l=1}^{N}\frac{1}{l^\beta}}\sum_{n=1}^{N} f_c(n) f_s(n)\frac{1}{n^\beta}\end{aligned} \tag{4.13}$$

From Eq. (4.13), we observe that larger transmission distance d_{ij} leads to a decrease of meeting probability (MP) ω_j. From this perspective, the paring range of i, TSZ $\mathcal{T}_i$, does not need to be too large. However, larger $\mathcal{T}_i$ may bring benefits to more receivers who need this file for a fixed receiver density. For an optimal TSZ of a transmitter in a general view, we then analyze the total number of receivers that successfully receive their required files, i.e. the BUE number N_r in $\mathcal{T}_i$. Let d_σ be an infinitely small area of the TSZ on interval $\left[\tau, \tau + d\tau\right]$. Since the probability that there are more than two UEs in d_σ can be neglected, the number of receivers that obtain required files in d_σ can be expressed as $\lambda_r \omega_{d_\sigma}$, where ω_{d_σ} is the receiving probability at d_σ and can be obtained from Eq. (4.13) through using τ instead of d_{ij}, i.e.

$$\omega_{d_\sigma} = \frac{\exp\left(-\kappa \tau^2\right)}{\sum_{l=1}^{N}\frac{1}{l^\beta}}\sum_{n=1}^{N} f_c(n) f_s(n)\frac{1}{n^\beta} \tag{4.14}$$

With the number of BUEs at location d_σ, N_r in the TSZ can be calculated by traversing all locations in the TSZ as follows:

$$\begin{aligned} N_r &= \int_{\mathcal{T}_i} \lambda_r \omega_{d_\sigma} \, d\sigma \\ &= \lambda_r \int_0^{2\pi} \int_0^{d_0} \omega_{d_\sigma} \tau \, d\tau \\ &= \frac{\pi \lambda_r}{\kappa \sum_{l=1}^{N} \frac{1}{l^\beta}} \left[1 - \exp\left(-\kappa d_0^2\right)\right] \sum_{n=1}^{N} f_c(n) f_s(n) \frac{1}{n^\beta}. \end{aligned} \tag{4.15}$$

Proposition 4.1 *Given a transmitter, the number N_r of BUEs in its TSZ increases with the increase in paring range of the TSZ, while this increase has an upper bound. Moreover, the upper bound N_r^{ub} of N_r can be expressed by:*

$$N_r^{ub} = \frac{\pi \lambda_r}{\kappa \sum_{l=1}^{N} \frac{1}{l^\beta}} \sum_{n=1}^{N} f_c(n) \, f_s(n) \frac{1}{n^\beta} \tag{4.16}$$

Proof: Taking the derivative of N_r in Eq. (4.15) with respect to d_0, we can obtain the derivation ν as follows:

$$\begin{aligned} \nu &= \frac{\partial}{\partial d_0} \frac{\pi \lambda_r}{\kappa \sum_{l=1}^{N} \frac{1}{l^\beta}} \left[1 - \exp\left(-\kappa d_0^2\right)\right] \sum_{n=1}^{N} f_c(n) f_s(n) \frac{1}{n^\beta} \\ &= 2\kappa d_0 \exp\left(-\kappa d_0^2\right) \frac{\pi \lambda_r}{\kappa \sum_{l=1}^{N} \frac{1}{l^\beta}} \sum_{n=1}^{N} f_c(n) f_s(n) \frac{1}{n^\beta} > 0 \end{aligned} \tag{4.17}$$

Thus, N_r increases with d_0, i.e. the paring range of the TSZ. Since

$$\lim_{d_0 \to \infty} \nu = 0 \tag{4.18}$$

N_r has an upper bound N_r^{ub}, which can be calculated by:

$$N_r^{\text{ub}} = \lim_{d_0 \to \infty} \frac{\pi \lambda_r}{\kappa \sum_{l=1}^{N} \frac{1}{l^\beta}} \left[1 - \exp\left(-\kappa d_0^2\right)\right] \sum_{n=1}^{N} f_c(n) f_s(n) \frac{1}{n^\beta} = \frac{\pi \lambda_r}{\kappa \sum_{l=1}^{N} \frac{1}{l^\beta}} \sum_{n=1}^{N} f_c(n) f_s(n) \frac{1}{n^\beta} \tag{4.19}$$

For a targeted TSZ with radius d_0, the transmission power needs to be $E_i = \eta_0 d_0^{k\alpha}$ according to the power control strategy. That is, the delivery range is determined by the transmission power as follows:

$$d_0 = \left(\frac{E_i}{\eta_0}\right)^{\frac{1}{k\alpha}} \tag{4.20}$$

Considering that d_0 increases with the transmission power, we can conclude from above analyses that N_r increases with the D2D transmission power before achieving the upper bound. However, it is not energy-efficient to increase N_r at the cost of increasing energy consumption due to the low energy capacity of UE and the energy conservation requirement of green networks.

To increase N_r with less energy consumption, we firstly evaluate the efficiency of content delivery in cellular networks. Denote the transmission efficiency coefficient ρ as the average energy consumption of a successful acquisition for required files at receivers in a TSZ, i.e.

$$\rho = \frac{E_i}{N_r} \tag{4.21}$$

Through applying Eq. (4.15) into (4.21), we can obtain ρ as follows:

$$\rho = \frac{E_i}{\frac{\pi\lambda_r}{\kappa\sum_{l=1}^{N}\frac{1}{l^\beta}}\left[1-\exp\left(-\kappa d_0^2\right)\right]\sum_{n=1}^{N} f_c(n) f_s(n)\frac{1}{n^\beta}} \tag{4.22}$$

Since the energy consumption and interference among transmitters also increase with E_i, there is a trade-off between E_i and N_r. Concerning the optimal transmission power and its corresponding TSZ of a given transmitter, we have the following proposition.

Proposition 4.2 *To minimize the transmission efficiency coefficient ρ under the transmission power constraint $E_{\max}$ of a UE, i.e. minimizing the average energy consumption for a successful file delivery from transmitter i to its receivers, the optimal transmission power E_i^* should satisfy*

$$E_i^* = \begin{cases} \widetilde{E}_i & \widetilde{E}_i \leq E_{\max} \\ E_{\max} & \widetilde{E}_i > E_{\max} \end{cases} \tag{4.23}$$

where $\widetilde{E}_i$ needs to satisfy

$$\exp\left(-\frac{\kappa}{\eta_0^{m'}}\widetilde{E}_i^{m'}\right) + \frac{\kappa m'\widetilde{E}_i^{m'}}{\eta_0^{m'}}\exp\left(-\frac{\kappa}{\eta_0^{m'}}\widetilde{E}_i^{m'}\right) - 1 = 0 \tag{4.24}$$

Proof: To minimize the average energy consumption of file delivery for a BUE, the optimal problem can be formulated as follows:

$$\min\ \rho = \frac{E_i}{\frac{\pi\lambda_r}{\kappa\sum_{l=1}^{N}\frac{1}{l^\beta}}\left[1-\exp\left(-\kappa d_0^2\right)\right]\sum_{n=1}^{N} f_c(n) f_s(n)\frac{1}{n^\beta}} \quad \text{s.t.} \quad 0 < E_i \leq E_{\max} \tag{4.25}$$

To solve (4.25), we include $d_0 = \left(\frac{E_i}{\eta_0}\right)^{\frac{1}{\kappa\alpha}}$ into the objective function (4.25), and then take the derivative of ρ with respect to E_i, i.e.

$$
\begin{aligned}
\nu' &= \frac{\partial}{\partial E_i} \frac{E_i}{\frac{\pi\lambda_c}{\kappa\sum_{l=1}^{N}\frac{1}{l^\beta}}\left[1-\exp\left(-\kappa d_0^2\right)\right]\sum_{n=1}^{N} f_c(n) f_s(n)\frac{1}{n^\beta}} \\
&= \left[\frac{\pi\lambda_r}{\kappa\sum_{l=1}^{N}\frac{1}{l^\beta}}\sum_{n=1}^{N} f_c(n) f_s(n)\frac{1}{n^\beta}\right]^{-1} \frac{\partial\frac{E_i}{\left[1-\exp\left(-\frac{\kappa}{\eta_0^{m'}}E_i^{m'}\right)\right]}}{\partial E_i} \\
&= \mathcal{A}\frac{\left[1-\exp\left(-\frac{\kappa}{\eta_0^{m'}}E_i^{m'}\right)\right]-\frac{\kappa m' E_i^{m'}}{\eta_0^{m'}}\exp\left(-\frac{\kappa}{\eta_0^{m'}}E_i^{m'}\right)}{\left[1-\exp\left(-\frac{\kappa}{\eta_0^{m'}}E_i^{m'}\right)\right]^2}
\end{aligned}
\tag{4.26}
$$

where $m' = \frac{2}{\kappa\alpha}$ and $\mathcal{A} = \left[\frac{\pi\lambda_r}{\kappa\sum_{l=1}^{N}\frac{1}{l^\beta}}\sum_{n=1}^{N} f_c(n) f_s(n)\frac{1}{n^\beta}\right]^{-1}$.

Letting $\nu' = 0$, we can obtain the optimal $\widetilde{E}_i$ satisfying

$$
\exp\left(-\frac{\kappa}{\eta_0^{m'}}\widetilde{E}_i^{m'}\right)+\frac{\kappa m'\widetilde{E}_i^{m'}}{\eta_0^{m'}}\exp\left(-\frac{\kappa}{\eta_0^{m'}}\widetilde{E}_i^{m'}\right)-1=0 \tag{4.27}
$$

which can be resolved easily using numerical methods.

Based on Proposition 4.2, the optimal radius d_0^* of a TSZ for a transmitter can be calculated by using (4.20), i.e. $d_0^* = \left(\frac{E_i^*}{\eta_0}\right)^{\frac{1}{\kappa\alpha}}$. In addition, from this proposition, we can find that E_i^* is determined by wireless environment parameters such as transmitter density, path loss factor, decoding threshold, and so on, and is independent to the caching distribution and content delivery distributions (f_c, f_s). However, f_c and f_s affect the probability that BUEs in the TSZ receive their required files. Without considering energy consumption, d_0 and N_r can increase infinitely with the transmission power. Although more UEs may benefit from once multicast, the benefit of D2D content delivery is reduced compared with BS content delivery, and interference becomes severer due to longer transmission distance.

4.3.2 Energy Efficiency for Multicast and Unicast

With the push strategy studied in this chapter, more than one BUE may exit for once multicast. To understand the advantages and constraints of energy efficiency for this push strategy, we compare the energy efficiency of D2D multicast and unicast (request-and-response strategy) here. For fair comparison, we evaluate the average transmission energy consumption per request for two strategies to show which one is more energy-efficient via taking content request, user distribution, and wireless transmission environment into consideration.

For the unicast content delivery, the transmission power can be expressed by:

$$E_i = \eta_0 d_{ij}^{\alpha} \tag{4.28}$$

For a constructed unicast link $i \to j$, the energy consumption of a successful delivery can be calculated by:

$$\varepsilon_u^{\text{succ}} = \frac{E_i}{\Pr\left\{\gamma_{ij} \geq \gamma_{\text{th}} | i\right\}} = \frac{\eta_0 d_{ij}^{\alpha}}{\exp\left(-\kappa d_{ij}^2\right)} \tag{4.29}$$

By averaging the location of transmitters, the average consumed energy of serving a request, ρ_u, can be calculated by:

$$\begin{aligned}\rho_u &= \int_S \varepsilon_u^{\text{succ}}\, d\sigma \\ &= \int_0^{2\pi}\int_0^{d_0} \lambda_t \frac{\eta_0 \tau^{\alpha}}{\exp\left(-\kappa\tau^2\right)} \tau\, d\tau \\ &= \lambda_t B\left(1, \frac{\alpha}{2}+1\right) d_{01}^2 F_1\left(\frac{\alpha}{2}+1; \frac{\alpha}{2}+2; \kappa d_0^2\right)\end{aligned} \tag{4.30}$$

where σ is an infinite small area, $B(x, y)$ is the beta function and ${}_1F_1(x; y; z)$ is the Degenerate hypergeometric function.

For the multicast content delivery, the transmission power is $\eta_0 d_0^{\alpha}$ for once transmission since the targeted delivery distance is d_0. Taking power control into consideration, unicast may be more energy-efficient than multicast. For example, although the farthest receiver in the targeted multicast range does not need the transmitted content, the transmission power is also fixed as ε_m. Even if the transmission power can be adaptively adjusted according to the distance of the farthest receiver requiring this content, energy is also consumed by overhead. However, multiple UEs may benefit from once multicast by receiving required or to be required contents. In this perspective, multicast is more energy-efficient than unicast. As analyzed in the last subsection, the average energy consumed by a successful content requirement for a BUE is expressed by ρ, which is re-written as:

$$\rho = E_i \left[\frac{\pi\lambda_r}{\kappa \sum_{l=1}^{N} \frac{1}{l^{\beta}}} \left[1 - \exp\left(-\kappa d_0^2\right)\right] \sum_{n=1}^{N} f_c(n) f_s(n) \frac{1}{n^{\beta}} \right]^{-1} \tag{4.31}$$

Hence, we can resort to (4.30) and (4.31) to compare the energy efficiency of two strategies. From the comparison, we observe that not only the UE densities (receiver density and transmitter density) but also caching and delivery strategies will affect the final efficiency of D2D multicast delivery.

4.3.3 Caching and Delivery

From Eq. (4.15), we can see that, besides the transmission power E_i, the number of BUEs that receive required files (N_r) also depends on the caching and delivery schemes, i.e. the probability for the transmitter to cache and disseminate files. In addition, they also impact the advantages of the push strategy. Here we discuss the caching and delivery distributions (f_c and f_s) of a transmitter to improve the D2D content delivery efficiency.

Let ζ denote a caching and delivery efficiency coefficient which characterizes the impact from caching and delivery distributions on the delivery performance, i.e.

$$\zeta = \sum_{n=1}^{N} f_c(n) f_s(n) \frac{1}{n^{\beta}} \tag{4.32}$$

The derivation of Eq. (4.15) shows that N_r grows with ζ for a given transmission power. Thus, to maximize N_r, i.e. maximizing the average number of receivers that obtain the required files, ζ needs to be maximized. The optimal problem can be formulated as follows:

$$\begin{aligned} \max \quad & \zeta = \sum_{n=1}^{N} f_c(n) f_s(n) \frac{1}{n^{\beta}} \\ \text{s.t.} \quad & \sum_{n=1}^{N} f_c(n) = 1 \\ & \sum_{n=1}^{N} f_s(n) = 1 \end{aligned} \tag{4.33}$$

With the Lagrange multipliers method, the objective function of this optimal problem can be written as:

$$\begin{aligned} \mathcal{L}(\mathbf{g}, \delta) = & \sum_{n=1}^{N} f_c(n) f_s(n) \frac{1}{n^{\beta}} + \delta_1 \left(\sum_{n=1}^{N} f_c(n) - 1 \right) \\ & + \delta_2 \left(\sum_{n=1}^{N} f_s(n) - 1 \right) \end{aligned} \tag{4.34}$$

where δ_1 and δ_2 are the Lagrange multipliers. To maximize the objective function in (4.33), we calculate the partial derivation of $\mathcal{L}(\mathbf{g}, \delta)$ with respect to $f_s(n)$ and $f_c(n)$ by:

$$\frac{\partial}{\partial f_c(n)} \mathcal{L}(\mathbf{g}, \delta) = f_s(n) \frac{1}{n^{\beta}} + \delta_1, \quad n = 1, 2, \ldots, N \tag{4.35a}$$

$$\frac{\partial}{\partial f_s(n)} \mathcal{L}(\mathbf{g}, \delta) = f_c(n) \frac{1}{n^{\beta}} + \delta_2, \quad n = 1, 2, \ldots, N \tag{4.35b}$$

Since for all files, i.e. $n = 1, 2, \ldots, N$, (4.35a) needs to be satisfied, we can obtain the relationship of the cached probabilities among these different files, which can be expressed as:

$$\frac{f_c(p)}{f_c(q)} = \left(\frac{q}{p} \right)^{\beta} \quad p \neq q \tag{4.36}$$

Due to $\sum_{n=1}^{N} f_c(n) = 1$, the distribution of f_c also follows the Zipf distribution, i.e.

$$f_c(n) = \frac{\frac{1}{n^{\beta}}}{\sum\limits_{l=1}^{N} \frac{1}{l^{\beta}}}, \quad n = 1, 2, \ldots, N \tag{4.37}$$

Similarly, from (4.35b) we can obtain the relationship of transmission probabilities among these different files, which can be expressed as follows:

$$\frac{f_s(p)}{f_s(q)} = \left(\frac{q}{p}\right)^{\beta}, \quad p \neq q \tag{4.38}$$

The distribution of f_s also follows Zipf distribution as the distribution of required files, i.e.

$$f_s(n) = \frac{\frac{1}{n^{\beta}}}{\sum_{l=1}^{N} \frac{1}{l^{\beta}}}, \quad n = 1, 2, \ldots N \tag{4.39}$$

Theorem 4.1 *To maximize the delivery efficiency, i.e. maximize the number of BUEs in the TSZ, the optimal file delivery distribution and the file caching distribution are Zipf when the required file distribution follows Zipf.*

Based on above theorem, more UEs can benefit from once multicast. The caching and delivery at UEs can be controlled by BSs to make UEs friendly and mutually beneficial.

4.4 D2D Delivery Mode Selection

The intention of D2D delivery is to improve resource efficiency for the cellular content delivery by using proactive cache and proximal transmission. However, this efficiency gain depends on many factors, such as UE density, file caching and delivery distributions at UEs, channel quality, and so on. In this section, we will compare D2D and BS content delivery in term of energy efficiency to understand the benefits and practical limits of the D2D delivery. Moreover, we discuss the criteria of mode selection for energy efficiency delivery, which is critical to integrate D2D into cellular networks [39–41]. Here we use the expected energy consumption for a successful file acquisition at a receiver as the evaluation metric. For a receiver j which requires a file, the probability $\xi_D(i)$ of receiving its required file, called MP, which can be written as:

$$\xi_D(i) = \sum_{n=1}^{N} \exp\left(-\kappa d_{ij}^2\right) f_s(n) f_c(n) f_r(n) \tag{4.40}$$

Then the expectation of $\xi_D(i)$, ξ_D, for a receiver under the D2D delivery mode can be calculated by averaging the location of i as follows:

$$\begin{aligned} \xi_D &= \int_0^{2\pi} \int_0^{d_0} \lambda_t \xi_D(i)\, \tau \, d\tau \\ &= 2\pi\lambda_t \int_0^{d_0} \frac{\exp\left(-\kappa\tau^2\right)}{\sum_{l=1}^{N} \frac{1}{l^{\beta}}} \sum_{n=1}^{N} f_s(n) f_c(n) \frac{1}{n^{\beta}} \tau \, d\tau \end{aligned}$$

$$= \frac{\pi\lambda_t\left[1 - \exp\left(-\kappa d_0^2\right)\right]}{\kappa \sum_{l=1}^{N} \frac{1}{l^\beta}} \sum_{n=1}^{N} f_s(n) f_c(n) \frac{1}{n^\beta} \tag{4.41}$$

With the BS content delivery, the probability that UE j receives the required file from its serving BS only depends on the channel status and delivery distribution under the assumption that BS always stores the required files of UEs. Then MP with the BS delivery can be expressed as:

$$\xi_C(i) = \sum_{n=1}^{N} \Pr\left\{\gamma_j \geq \gamma_0\right\} f_s(n) f_r(n) \tag{4.42}$$

where

$$\Pr\left\{\gamma_j \geq \gamma_{\text{th}}\right\} = \exp\left(-\kappa' d_{ij}^2\right) \tag{4.43}$$

and d_R is the coverage radius of the BS with $\kappa' = \pi\lambda_b m\Gamma(m)\Gamma(1-m)\gamma_{\text{th}}^m$. The derivation of (4.43) is similar to that of (4.8) and the difference is that the BS density is λ_b. By averaging locations of all potential serving BSs of the UE j, the expectation of $\xi_C(i)$, ξ_C, can be written as:

$$\xi_C = \frac{\sum_{n=1}^{N} f_s(n) \frac{1}{n^\beta}}{\sum_{l=1}^{N} \frac{1}{l^\beta}} \int_0^{d_R} \exp\left(-\kappa'\tau^2\right) f(\tau)\, d\tau \tag{4.44}$$

where $f(\tau)$ is the distribution of the distance between a receiver and its serving BS. In cellular networks, a UE usually selects its attached BS based on received signal strength indicator (RSSI). Hence, the distribution of the distance between a UE and its serving BS $f(\tau)$ can be expressed as:

$$f(\tau) = 2\pi\lambda_b\tau \exp\left(-\lambda_b\pi\tau^2\right) \tag{4.45}$$

Including (4.45) into (4.44), ξ_C can be written as:

$$\begin{aligned} \xi_C &= \frac{\sum_{n=1}^{N} f_s(n) \frac{1}{n^\beta}}{\sum_{l=1}^{N} \frac{1}{l^\beta}} \int_0^{d_R} \exp\left(-\kappa'\tau^2\right) 2\pi\lambda_b\tau \exp\left(-\lambda_b\pi\tau^2\right) d\tau \\ &= \frac{\sum_{n=1}^{N} f_s(n) \frac{1}{n^\beta}}{\sum_{l=1}^{N} \frac{1}{l^\beta}} \frac{\pi\lambda_b}{\kappa' + \pi\lambda_b} \left[1 - \exp\left(\left(-\kappa' - \pi\lambda_b\right) d_R^2\right)\right] \end{aligned} \tag{4.46}$$

For a fair comparison between cellular and D2D delivery in term of energy efficiency, we compare the MPs under different modes with the same transmission energy and delivery coverage. Let the ratio between ξ_C and ξ_D be ε_e. With (4.41) and (4.46), we can obtain ε_e as follows:

Table 4.1 Mode selection algorithm for content delivery.

1. For a D2D UE i scheduled to transmit in a content delivery period
2. BS calculates ε_e based on (4.47)
3. If $\varepsilon_e \leq 1$,
4. UE i transmits cached file n with probability $f_s(n)$
5. BS schedules UEs within $\mathcal{T}_i$ to listen to the multicast file
6. else
7. BS transmits cached file n with probability $f_s(n)$
8. BS schedules attached UEs to listen to the multicast file
9. end

$$
\begin{aligned}
\varepsilon_e &= \frac{\xi_C}{\xi_D} \\
&= \frac{\sum_{n=1}^{N} \frac{f_s(n)}{n^\beta} \left[\sum_{l=1}^{N} \frac{1}{l^\beta}\right]^{-1} \frac{\pi\lambda_b}{\kappa'+\pi\lambda_b} [1-\exp((-\kappa'-\pi\lambda_b)d_R^2)]}{\pi\lambda_t[1-\exp(-\kappa d_0^2)] \left[\kappa \sum_{l=1}^{N} \frac{1}{l^\beta}\right]^{-1} \sum_{n=1}^{N} f_s(n) f_c(n) \frac{1}{n^\beta}} \\
&= \frac{\kappa\lambda_b \{1-\exp[(-\kappa'-\pi\lambda_b)d_R^2]\} \sum_{n=1}^{N} f_s(n) \frac{1}{n^\beta}}{\lambda_t(\kappa'+\pi\lambda_b)[1-\exp(-\kappa d_0^2)] \sum_{n=1}^{N} f_s(n) f_c(n) \frac{1}{n^\beta}}
\end{aligned}
\tag{4.47}
$$

From (4.47), we can find that ε_e is affected by multiple factors such as the density of BS, D2D transmitters, the paring range, the distributions of delivery and caching. The transmission mode of a UE is determined by its serving BS based on ε_e. That is, BS selects the cellular mode when $\varepsilon_e > 1$. Otherwise, BS selects the D2D mode. The detailed mode selection algorithm is listed in Table 4.1. Here we can see that the mode selection is based on the spatial average performance of communication links which is independent of the location of a transmitter. Then all transmitters will perform the same mode selection decision to achieve a better average performance without much signaling overhead cost. Thus, the distribution of transmitters still follows homogenous PPP after mode selection.

4.5 Performance Evaluation

To evaluate proposed schemes, we present some numerical and simulation results in this section. The default parameters are listed in Table 4.2. In the simulation, each UE has a single role, i.e. transmitter or receiver during a simulation round. The network performance can be well reflected due to the randomness of their locations. More detailed simulation and relating parameters setup for D2D communication can be referred to [34]. Due to different system model and problem formulation, we will not compare with existing works such as [42] in the simulation. Instead, we widely simulate various transmission environment parameters to check their

Table 4.2 Simulation parameter settings.

Simulation parameter	Value
Transmitter UE density λ_t	10^{-7}, 10^{-6}, 10^{-5}
BS density λ_b	10^{-7}, 10^{-6}, 10^{-5}
Receiver UE density λ_r	10^{-2}
SINR threshold γ_{th}	10 dB
Targeted receiving power η_0	−80 dBm
Maximum transmission power $E_{\max}$	31 dBm
Path loss compensatory factor k	1
Path loss factor α	3
Zipf exponent β	2
Files number N	100
Simulation times T	1000

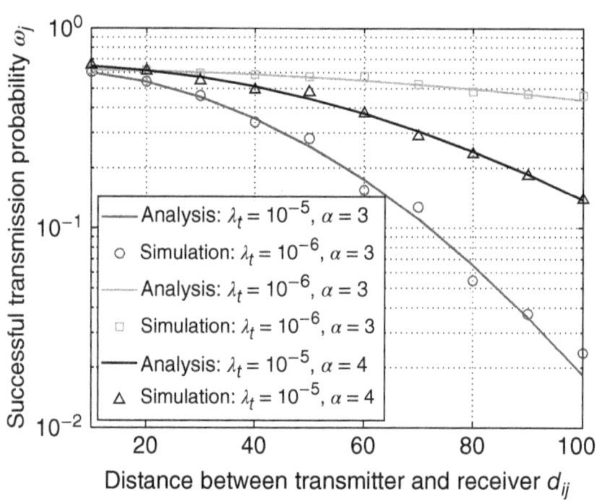

Figure 4.1 Successfully receiving probability ω_j against the D2D link distance d_{ij}.

impacts on D2D multicast delivery. In addition, we indicate that the proposed D2D multicast strategy achieve the best energy efficiency coefficient by simulation.

Firstly, we compare successfully receiving probability (ω_j) at a receiver against the distance between receiver and transmitter (d_{ij}) under different settings of transmitter densities (λ_t) and path loss factors (α) in Figure 4.1. We can find that ω_j decreases with d_{ij} under different λ_t and α. Thus, a larger TSZ (leading to larger link distances for more transmissions) may not increase N_r. Moreover, comparing lines with circles and squares, respectively, we can observe that the higher transmitter density leads to a smaller successfully receiving probability. Larger α leads to larger ω_j at the higher transmitter density, say $\lambda_t = 10^{-5}$, since interference from simultaneous transmitters decreases with α.

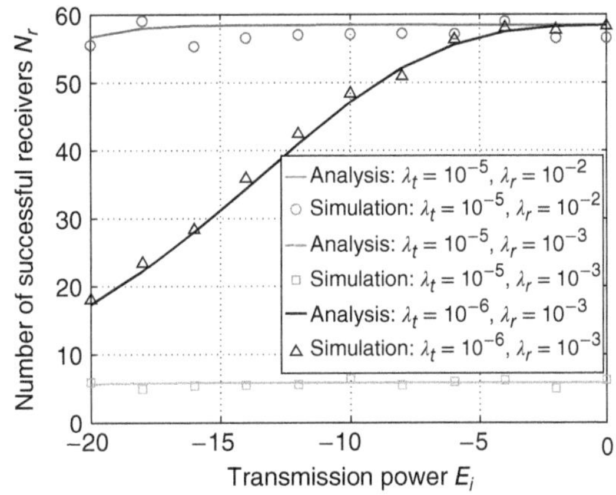

Figure 4.2 The number of UEs successfully receiving their required files N_r against transmission power E_i.

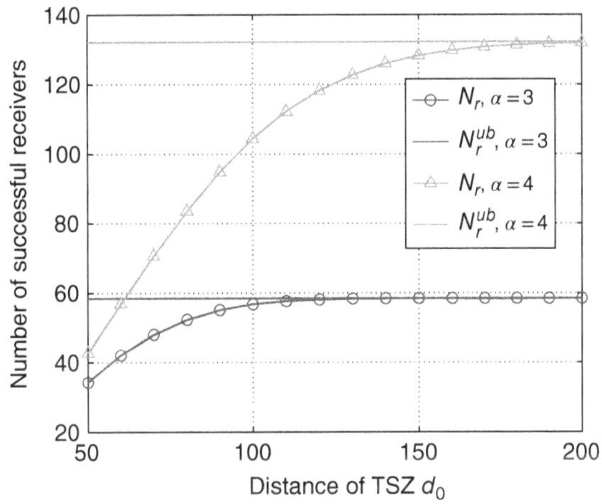

Figure 4.3 The number of UEs successfully receiving their required files N_r and corresponding upper bound N_r^{ub} against radii of TSZ d_0.

As shown in Figure 4.2 for N_r against d_0, N_r increases with the radius of TSZ (d_0). However, this increase is trivial for the higher transmitter density (say $\lambda_t = 10^{-5}$) due to interference. In particular, for the lower receiver density, N_r increases more slowly. Thus, the transmission power needs to be adjusted based on nodes density to avoid energy waste.

Figure 4.3 compares N_r and N_r^{ub} against d_0 for $\lambda_t = 10^{-5}$. As shown in this figure, N_r increases with d_0. However, N_r tends to be a constant when d_0 is larger than a threshold, i.e. N_r has an upper bound as analyzed. In addition, N_r tends to reach the upper bound more quickly with smaller α (say $\alpha = 3$) due to severer interference from simultaneous transmitters, which match the theoretical analyses, i.e. it is not energy-efficient to increase N_r by enlarging the TSZ of a transmitter.

As shown in Figure 4.4, the energy consumption (E_i) increases with d_0 under different α and η_0. In addition, higher η_0 and α lead to higher energy consumptions for the same d_0, because they need higher transmitter powers to achieve the targeted receiving power and compensate path loss, respectively. Thus, d_0 needs to be reduced to save energy. Figure 4.3 also shows that N_r increases with d_0 with an upper

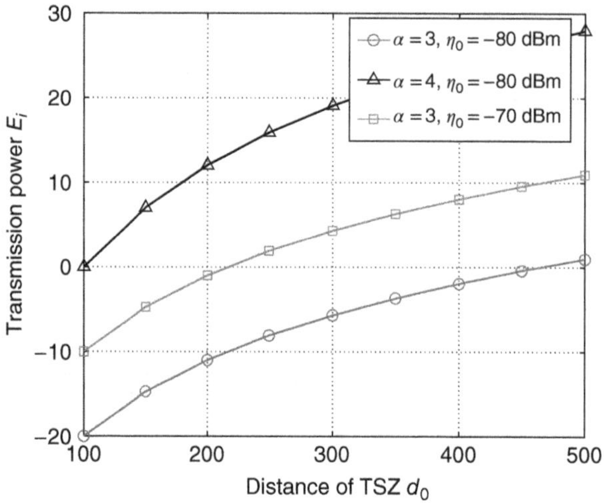

Figure 4.4 The transmission power E_i against radius of TSZ d_0 under different path loss factor α and targeted receiving signal power η_0.

bound, i.e. N_r does not increase any more when d_0 is larger than a threshold. Thus, an optimal d_0 exists especially for a worse transmission environment as analyzed.

The comparisons of transmission efficiency coefficient (ρ) with different settings of d_0, α, and γ_{th} are shown in Figure 4.5. It is shown that ρ firstly decreases with the increase of d_0 for small d_0 with ranging from 0 to 50 and then increases with d_0 after achieving its minimal value. Hence, there is an optimal d_0 for the TSZ given transmission environment parameters such as transmitter density, receiver density, and channel quality. In addition, the optimal ρ can achieve the minimal value under different transmission environments, which verifies the effectiveness of the proposed energy efficiency-oriented D2D delivery scheme. Moreover, we can find that ρ increases with path loss and decoding threshold, while the best energy efficiency can always be achieved by the proposed delivery power scheme.

Similarly as shown in Figure 4.6, ρ of the proposed scheme (marked by star points) can obtain the optimal energy efficiency coefficient at different transmitter densities with the smallest transmission power consumption for the same number of successful decoding receivers. In addition, a higher transmitter density yields a higher ρ, i.e. a lower energy efficiency of the D2D delivery due to severer interference.

Figure 4.7 compares ρ with different d_0 and path loss compensation factor k. We can see that ρ increases with the increase of k, since larger path loss leads to a smaller number of successful receiving UEs and the lower energy efficiency.

The above discussions show that, under different transmission environments (path loss, transmitter density, and decoding threshold), the proposed delivery scheme can achieve the best energy efficiency.

For fair comparison among different caching and delivery strategies, the number of BUEs (N_r) should be compared for different schemes. To focus on caching and delivery strategies, communication parameters such as transmission power and channel quality are set the same for different schemes. From Eq. (4.15), we can see

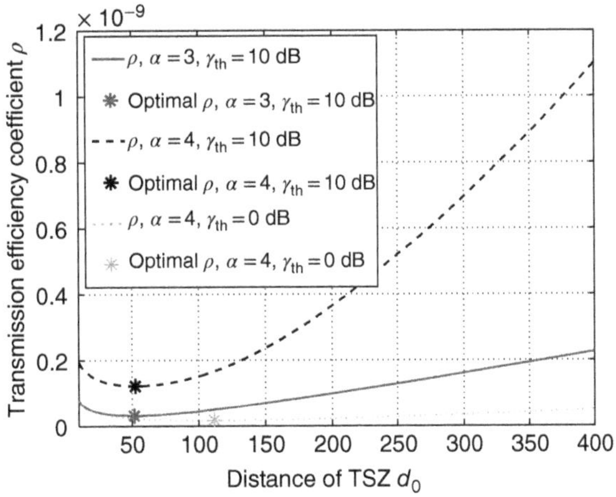

Figure 4.5 The transmission efficiency coefficient ρ with different radii of TSZ d_0, path loss factor α and decoding threshold γ_{th}.

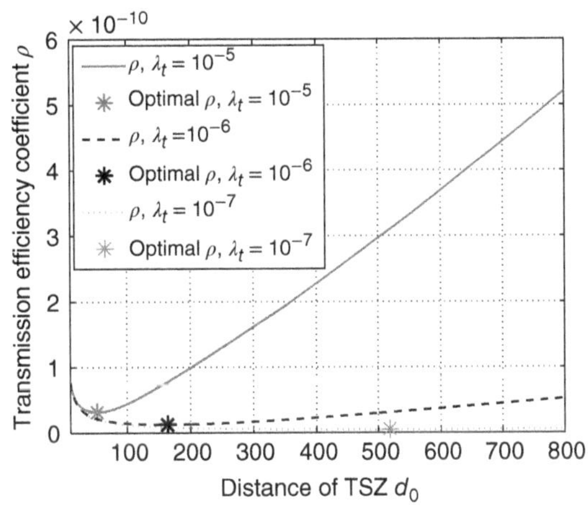

Figure 4.6 The transmission efficiency coefficient ρ with different TSZ radii d_0 and transmitter density λ_t.

that N_r only depends on ζ when communication parameters are given. Therefore, we can resort to comparing ζ to determine which scheme is more energy-efficient. Figure 4.8 compares ζ with different files number (N) for three different content caching and delivery schemes: random caching and delivery, Zipf caching and random delivery, and the proposed optimal caching and delivery. We can find that the proposed scheme achieves the highest ζ (approaching to 1) as analyzed under any tested N, which will bring a largest number of BUEs from the delivery of a transmitter. Zipf caching is better than random caching with the random D2D delivery policy, and random caching and delivery performs worst especially with larger N: its performance degrades significantly with N. On the other hand, the proposed optimal scheme almost performs the same when N changes, which shows the robustness of the proposed scheme.

Figure 4.9 shows the variation of expectation of MP (ξ_D) under the D2D mode against λ_t. ξ_D increases with the increase of λ_t, since more transmitters have larger probability to pair with a receiver and send the required file. In addition, the increase

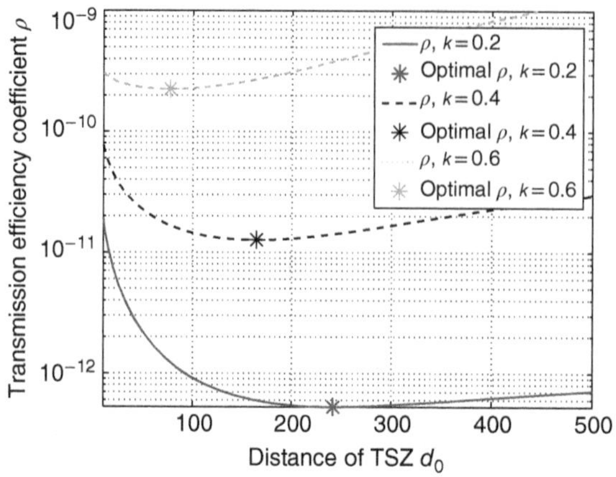

Figure 4.7 The transmission efficiency coefficient ρ with different TSZ radius d_0 and path loss compensation factor k.

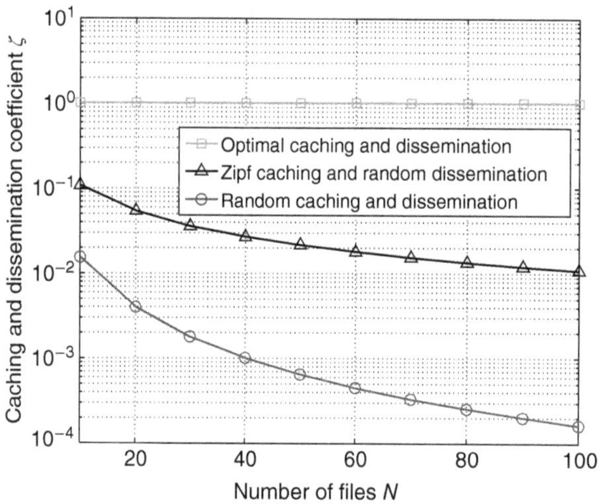

Figure 4.8 Caching and delivery coefficient ζ with different files number N.

of ξ_D with λ_t tends to zero after λ_t goes beyond a threshold due to the severer interference introduced by more transmitters. ξ_D decreases with the α and γ_{th} due to worse file receiving condition.

Figure 4.10 shows the expectation of MP (ξ_C) under the cellular mode against λ_b. The targeted coverage distance of the BS, d_R, is set equal to d_0 for better comparison. Similar to Figure 4.9, ξ_C also increases with λ_b with an upper bound due to interference from simultaneous transmitters. In addition, ξ_C decreases with the increase of α and γ_{th} due to worse file receiving condition. Through comparing Figures 4.9 and 4.10, we can find that under the same transmission environment, the BS delivery performs better than the D2D delivery, because BS stores all files instead of caching files opportunistically as the D2D UE does. Therefore, the D2D file delivery can be a complement choice under certain conditions.

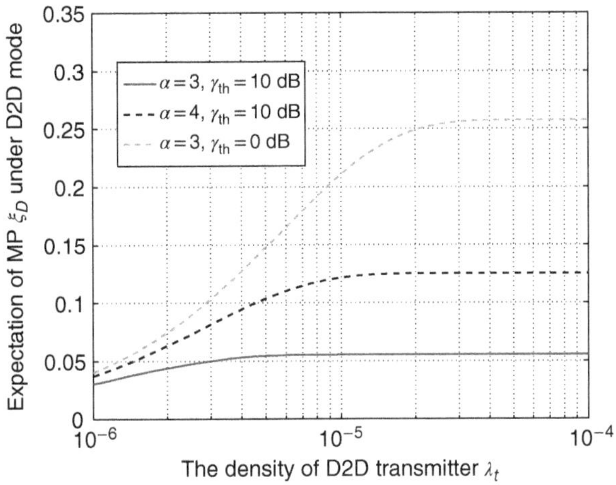

Figure 4.9 The expectation of MP ξ_D under D2D mode against D2D transmitter density λ_t.

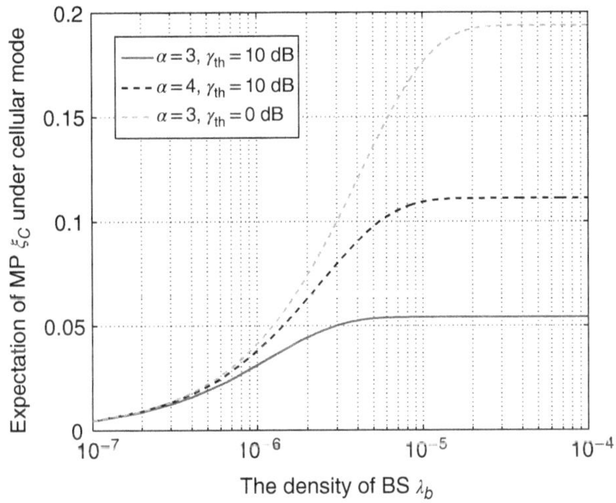

Figure 4.10 The expectation of MP ξ_C under cellular mode against BS density λ_b.

In Figure 4.11, ε_e against d_0 are compared for different settings of λ_b, β, and delivery strategies. The targeted delivery radii are set equal for the D2D transmitter and BS for fair comparison in term of energy efficiency. For the random delivery strategy, in which a D2D transmitter randomly caches files, we can observe that the D2D content delivery performs worse than the BS content delivery ($\varepsilon_e > 1$). However, for the optimal delivery strategy in which D2D transmitter caches and transmits files optimally, the performance of D2D content delivery is improved. When the D2D transmitter density is much larger than BS density (say $\lambda_t = 5\lambda_b$), the D2D delivery may outperform the BS delivery ($\varepsilon_e < 1$), since D2D transmitters bring content sharing closer to receivers due to the dense UE deployment. Thus, it is better to select the D2D delivery when available UEs for content multicast delivery are of higher

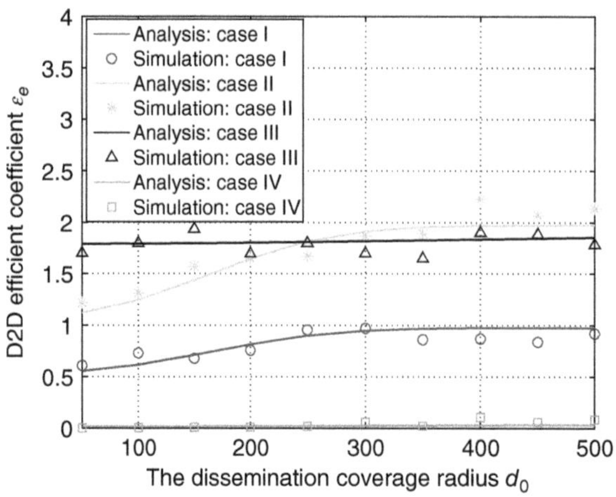

Figure 4.11 D2D energy-efficient coefficient ε_e under different cases. Case I: optimal delivery, $\beta = 2$ and $\lambda_t/\lambda_b = 2$, Case II: optimal delivery, $\beta = 0.5$ and $\lambda_t/\lambda_b = 2$, Case III: random delivery, $\beta = 0.5$ and $\lambda_t/\lambda_b = 2$, and Case IV: optimal delivery, $\beta = 0.5$ and $\lambda_t/\lambda_b = 5$.

density. For smaller β, the D2D delivery is worse than BS delivery due to limited caching capacity with not enough required files. Thus, D2D content delivery may perform better under some conditions. We can resort to ε_e to determine when to use the D2D delivery.

4.6 Summary

In this chapter, the D2D users are paired according to the optimal paring range (D2D multicast range), which maximizes the number of benefited users while save transmission energy. Then, we discuss an energy-efficient content delivery policy, which allows a transmitter to organize the delivery and caching of files to minimize the energy consumption for a successful file delivery. With the minimized transmission power, the interference of content delivery can be minimized. Lastly, we give a quantitative metric to evaluate the energy efficiency of D2D content delivery, which provides an important insight into determining whether D2D paring and corresponding content delivery should be used.

References

1 Bastug, E., Bennis, M., Kountouris, M., and Debbah, M.S. (2015). Cache-enabled small cell networks: modeling and tradeoffs. *EURASIP Journal on Wireless Communications and Networking* (1): 1–11.

2 Hoydis, J., Kobayashi, M., and Debbah, M. (2011). Green small-cell networks. *IEEE Vehicular Technology Magazine* 6 (1): 37–43.

3 Bastug, E., Guenego, J.-L., and Debbah, M. (2013). Proactive small cell networks. *Proceedings of 2013 20th International Conference on Telecommunications (ICT)*, pp. 1–5.

4 Shanmugam, K., Golrezaei, N., Dimakis, A.G. et al. (2013). FemtoCaching: wireless content delivery through distributed caching helpers. *IEEE Transactions on Information Theory* 59 (12): 8402–8413.

5 Golrezaei, N., Dimakis, A.G., and Molisch, A.F. (2014). Scaling behavior for device-to-device communications with distributed caching. *IEEE Transactions on Information Theory* 60 (7): 4286–4298.

6 Fodor, G., Dahlman, E., Mildh, G., and Parkvall, S. (2012). Design aspects of network assisted device-to-device communications. *IEEE Communications Magazine* 50 (3): 170–177.

7 Asadi, A., Wang, Q., and Mancuso, V. (2014). A survey on device-to-device communication in cellular networks. *IEEE Communications Surveys and Tutorials* 16 (4): 1801–1819.

8 Altieri, A., Piantanida, P., Vega, L.R., and Galarza, C.G. (2014). A stochastic geometry approach to distributed caching in large wireless networks. *Proceedings of 2014 11th International Symposium on Wireless Communications Systems (ISWCS)*, pp. 863–867.

9 Golrezaei, N., Mansourifard, P., Molisch, A.F., and Dimakis, A.G. (2014). Base-station assisted device-to-device communications for high-throughput wireless video networks. *IEEE Transactions on Wireless Communications* 13 (7): 3665–3676.

10 Paakkonen, J., Hollanti, C., and Tirkkonen, O. (2013). Device-to-device data storage for mobile cellular systems. *Proceedings of 2013 IEEE GLOBECOM Workshops*, pp. 671–676.

11 Haenggi, M., Andrews, J.G., Baccelli, F. et al. (2009). Stochastic geometry and random graphs for the analysis and design of wireless networks. *IEEE Journal on Selected Areas in Communications* 27 (7): 1029–1046.

12 Giatsoglou, N., Ntontin, K., Kartsakli, E. et al. (2017). D2D-aware device caching in mmWave-cellular networks. *IEEE Journal on Selected Areas in Communications* 35 (9): 2025–2037. https://doi.org/10.1109/JSAC.2017.2720818.

13 Malak, D. and Al-Shalash, M. (2014). Optimal caching for device-to-device content distribution in 5G networks. *Proceedings of 2014 GLOBECOM Workshops*, pp. 863–868.

14 Kang, H.J., Park, K.Y., Cho, K., and Kang, C.G. (2014). Mobile caching policies for device-to-device (D2D) content delivery networking. *Proceedings of 2014 IEEE Conference on Computer Communications Workshops*, pp. 299–304.

15 Zheng, Z., Wang, T., Song, L. et al. (2014). Social-aware multi-file dissemination in device-to-device overlay networks. *Proceedings of 2014 IEEE Conference on Computer Communications Workshops*, pp. 219–220.

16 Pyattaev, A., Galinina, O., Andreev, S. et al. (2015). Understanding practical limitations of network coding for assisted proximate communication. *IEEE Journal in Selected Areas of Communications* 33 (2): 156–170.

17 Bastug, E., Bennis, M., and Debbah, M. (2014). Living on the edge: the role of proactive caching in 5G wireless networks. *IEEE Transactions on Mobile Computing* 52 (8): 82–89.

18 Cao, Y. and Maaref, A. (2014). Device cooperation-assisted scalable video multicast with heterogeneous QoE guarantees. *Proceedings of 2014 11th International Symposium on Wireless Communications Systems (ISWCS)*, pp. 733–738, August 2014.

19 Zhang, Y., Li, F., Ma, X. et al. (2016). Cooperative energy-efficient content dissemination using coalition formation game over device-to-device communications. *Canadian Journal of Electrical and Computer Engineering* 39 (1): 2–10.

20 Xu, Y. and Liu, F. (2017). Qos provisionings for device-to-device content delivery in cellular networks. *IEEE Transactions on Multimedia.* 19 (11): 2597–2608.

21 Lin, X., Ratasuk, R., Ghosh, A., and Andrews, J.G. (2014). Modeling, analysis, and optimization of multicast device-to-device transmissions. *IEEE Transactions on Wireless Communications* 13 (8): 4346–4359.

22 Afshang, M., Dhillon, H.S., and Chong, P.H.J. (2016). Fundamentals of cluster-centric content placement in cache-enabled device-to-device networks. *IEEE Transactions on Communications* 64 (6): 2511–2526. https://doi.org/10.1109/TCOMM.2016.2554547.

23 Xu, Y., Li, X., and Zhang, J. (2017). Device-to-device content delivery in cellular networks: multicast or unicast. *IEEE Transactions on Vehicular Technology* (99): 1. https://doi.org/10.1109/TVT.2017.2787723.

24 Shalmashi, S., Björnson, E., Kountouris, M. et al. (2015). Energy efficiency and sum rate when massive MIMO meets device-to-device communication. *2015 IEEE International Conference on Communication Workshop (ICCW)*, pp. 627–632, June 2015. https://doi.org/10.1109/ICCW.2015.7247251.

25 Antonopoulos, A. and Verikoukis, C. (2014). Multi-player game theoretic MAC strategies for energy efficient data dissemination. *IEEE Transactions on Wireless Communications* 13 (2): 592–603.

26 Antonopoulos, A., Kartsakli, E., and Verikoukis, C. (2014). Game theoretic D2D content dissemination in 4G cellular networks. *IEEE Communications Magazine* 52 (6): 125–132.

27 Datsika, E., Antonopoulos, A., Zorba, N., and Verikoukis, C. (2017). Green cooperative device-to-device communication: a social-aware perspective. *IEEE Access* 4: 3697–3707.

28 Xu, C., Song, L., Han, Z. et al. (2012). Interference-aware resource allocation for device-to-device communications as an underlay using sequential second price auction. *2012 IEEE International Conference on Communications (ICC)*, pp. 445–449, June 2012.

29 Tsolkas, D., Liotou, E., Passas, N., and Merakos, L. (2012). A graph-coloring secondary resource allocation for D2D communications in LTE networks. *2012 IEEE 17th International Workshop on Computer Aided Modeling and Design of Communication Links and Networks (CAMAD)*, pp. 56–60, September 2012. https://doi.org/10.1109/CAMAD.2012.6335378.

30 Xu, Y., Liu, Y., and Li, D. (2015). Resource management for interference mitigation in device-to-device communication. *IET Communications* 9 (9): 1199–1207.

31 Zhou, Z., Dong, M., Ota, K. et al. (2014). Energy efficiency and spectral efficiency tradeoff in device-to-device (D2D) communications. *IEEE Wireless Communications Letters* 3 (5): 485–488.

32 Afshang, M., Dhillon, H.S., and Chong, P.H.J. (2016). k-Closest coverage probability and area spectral efficiency in clustered D2D networks. *2016 IEEE International Conference on Communications (ICC)*, pp. 1–6, May 2016. https://doi.org/10.1109/ICC.2016.7510857.

33 Al Haija, A.A. and Vu, M. (2015). Spectral efficiency and outage performance for hybrid D2D-infrastructure uplink cooperation. *IEEE Transactions on Wireless Communications* 14 (3): 1183–1198. https://doi.org/10.1109/TWC.2014.2365171.

34 3GPP WG RAN1 (2014). *3gpp Technical Report 36.843*. http://www.3gpp .\LY1\textbackslash\LY1\textbackslashorg/ftp/Specs/html-info/36843.htm (accessed 7 October 2021).

35 Golrezaei, N., Dimakis, A.G., and Molisch, A.F. (2012). Wireless device-to-device communications with distributed caching. *Proceedings of 2012 IEEE International Symposium on Information Theory Proceedings (ISIT)*, pp. 2781–2785.

36 Bastug, E., Bennis, M., and Debbah, M. (2014). Social and spatial proactive caching for mobile data offloading. *Proceedings of 2014 IEEE International Conference on Communications Workshops*, pp. 581–586.

37 Poularakis, K. and Tassiulas, L. (2014). Optimal selfishness-aware device-assisted content delivery in cellular networks. *Proceedings of 2014 IEEE Wireless Communications and Networking Conference (WCNC)*, pp. 2288–2293.

38 Xu, Y. (2016). On the performance of device-to-device communications with delay constraint. *IEEE Transactions on Vehicular Technology* 65 (11): 9330–9344. https://doi.org/10.1109/TVT.2016.2519456.

39 Elsawy, H., Hossain, E., and Alouini, M.S. (2014). Analytical modeling of mode selection and power control for underlay D2D communication in cellular networks. *IEEE Transactions on Communications* 62 (11): 4147–4161.

40 Sakr, A.H. and Hossain, E. (2015). Cognitive and energy harvesting-based D2D communication in cellular networks: stochastic geometry modeling and analysis. *IEEE Transactions on Communications* 63 (5): 1867–1880.

41 Lin, X., Andrews, J.G., and Ghosh, A. (2014). Spectrum sharing for device-to-device communication in cellular networks. *IEEE Transactions on Wireless Communications* 13 (12): 6727–6740.

42 Li, Y., Wang, Z., Jin, D., and Chen, S. (2014). Optimal mobile content downloading in device-to-device communication underlaying cellular networks. *IEEE Transactions on Wireless Communications* 13 (7): 3596–3608. https://doi.org/10.1109/TWC.2014.2315807.

5

Resource Allocation for NOMA-based D2D Systems Coexisting with Cellular Networks

Tien H. Nguyen[1], *Taehyun Yoon*[2], *Xuan T. Nguyen*[1], *Daeseung Yoo*[2], *Byungtae Jang*[2], *and Van D. Nguyen*[1]

[1] *School of Electronics and Telecommunications, Hanoi University of Science and Technology (HUST), Hanoi, Vietnam*
[2] *Electronics and Telecommunications Research Institute (ETRI), Daejeon, South Korea*

Device-to-device (D2D) communication is a promising technology that can alleviate traffic in core networks as well as increase mobility without rerouting data through NodeB [1]. Operating at short transmission distances with a low transmit power between D2D pairs, the spectral efficiency (SE) and energy efficiency (EE) can be significantly improved by sharing spectrum with cellular users (CUs). Traditionally, spectrum sharing is considered in the uplink because the downlink is often more crowded; however, the uplink and downlink loads have become more similar in modern high-density mobile networks [2–4]. D2D communications can potentially facilitate peer-to-peer (P2P) services [5, 6]. Motivated by these possibilities, many publications have studied different scenarios.

Nonorthogonal multiplexing access (NOMA) technology is capable of handling the demands of high throughput traffic in the next-generation networks [7]. NOMA offers significant improvements in both SE and EE without requiring any additional infrastructure manage resources [8, 9]. Although conventional orthogonal multiplexing access (OMA) remains a core technology to achieve high throughput performance [10] toward the fifth-generation (5G) goals of achieving higher network capacity, the development of 5G wireless networks poses new challenges by increasing the density of the entire network. This leads to an increased demand for NOMA development in both research and industry [11]. NOMA technology allows many users to share the same channel by varying the transmit power levels [12]. Successive interference cancelation (SIC) is then utilized at the receiver to process and sort the desired signals [13, 14]. Considering the potentials of D2D and NOMA, it is essential to explore whether NOMA can be combined with D2D communication systems to maximize efficiency.

Interference Mitigation in Device-to-Device Communications, First Edition.
Edited by Masood Ur Rehman, Ghazanfar Ali Safdar, and Mohammad Asad Rehman Chaudhry.

5.1 NOMA-based D2D Systems

While 5G will be standardized in Release 15, NOMA is positioned to be introduced in 6G. Many studies have been initiated worldwide to identify the potential benefits of NOMA. The first downlink NOMA transmission scenario was studied [15] using only two single-antenna users and multiple transmit antennas at the base station (BS). Later studies developed a multiple antenna BS that transmits to multiple users in the same subchannel using the power coding technique [16–19]. The challenges and advantages of NOMA are discussed elsewhere [12, 20]. A D2D network employing NOMA was previously considered [10], in which the authors proposed a novel concept to group the users in a NOMA-based D2D system to groups and multiple D2D groups were allowed to share subchannels. A similar study [21] proposed the joint allocation of subchannels and powers to maximize the total rate of the D2D-NOMA system, while satisfying the signal-to-interference-and-noise ratio (SINR) requirements of all D2D users in the network. The authors sequentially modeled the nonconvex power allocation to a convex problem. These concepts and grouping methods were extended to allow user pairing within groups before using NOMA to decode the desired signals between each pair [22, 23].

The impacts of user pairing was previously studied [24–27] in a centralized NOMA-D2D system where the BS controls both the sharing resources and manages the interference between cellular and D2D users. However, in a disjointed multiple user, multiple group NOMA uplink scenario, the interference between users at a particular BS increased proportionally with the number of neighboring transmitting users [6, 28–33]. To mitigate cluster interference, additional studies introduced beamforming-aided NOMA [34, 35]. Currently, work using use zero-force (ZF) or maximum ratio transmission (MRT) detection at the receiver has helped reduce intercluster interference in both perfect and imperfect channel state information (CSI) conditions [9, 11, 36, 37]. Closed-form expression investigating the ergodic sum rate for uplinks and downlink is discussed elsewhere [38, 39], respectively. Because the joint optimization problems in such models utilize high-dimensional semi-definite programming, contributors are utilizing convex optimization problems that can be solved in a heuristic way.

In this chapter, we consider coexistence between the NOMA-D2D downlink system and the uplink cellular system extending the centralized models developed in literature [24–27]. This scenario can occur when cellular users use D2D communication in a femtocell network, which is in the same geographic area of a cellular microcell network. Methods to reduce the interference in cellular systems, such as beamforming, [34, 35] are not suitable for D2D users because of the high cost of hardware as well as the introduction of unnecessary complexity to already dense networks. A comprehensive framework is presented to overcome the joint subchannel allocation, user pairing and power control in NOMA-D2D downlinks underlying a cellular uplink network. In contrast to previous studies, because the channels are assumed to have unique frequencies, each signal received at the SIC receiver experiences a distinct channel gain that includes path loss and shadowing. By exploiting these channel gains, we have proposed an effective subchannel allocation and user pairing method to mitigate the interference from cellular users to D2D users. This chapter describes our efforts to minimize the total transmit power of all devices in the network, while minimizing SINR requirements for both cellular users (CU) and D2D users (DU).

5.2 System Model and Performance Analysis

This section presents a detailed model of the NOMA-based system. Vectors and matrices are denoted by bold face small and big letters, correspondingly. The superscripts T and H stand for the transpose and conjugate transpose. $\mathbf{I}_K$ is the $K \times K$ identity matrix. The operator $\mathbb{E}\{.\}$ is the expectation of a random variable. The notation $\|.\|$ for the Euclidean norm. $\mathcal{CN}(\cdot,\cdot)$ stands for circularly complex Gaussian distribution. $\mathcal{O}$ is the complexity order of a process.

5.2.1 System Model and Assumptions

Our model utilizes a cellular BS equipped with N antennas and schedules L orthogonal subchannels. In addition, there are M downlink D2D-based NOMA D2D groups that transmit in the same geographical area. All D2D devices (i.e. D2D transmitter and D2D users) have a single antenna. In contrast to traditional D2D communications pairs [40, 41], the users in each D2D group will be paired such that SIC at the receiver side can be combined with NOMA, as shown in Figure 5.1. k^{th}, $k \in K_m$ D2D receiver pairs, denoted as DR_m, will simultaneously receive a pair of messages, s_k and $s_{k'}$ from DT_m D2D transmitters. In general, all channels in the network follow Rayleigh distribution patterns but can be classified into two groups:

- *Uplink channels between cellular users, D2D transmitter devices, and cellular BS*: The column vector $\mathbf{h}_{l,n} \in \mathbb{C}^N$ denotes the channel between CU n that uses subchannel l and cellular BS; The column vector $\mathbf{g}_m \in \mathbb{C}^N$ denotes the channel between the D2D transmitter in group m and cellular BS.

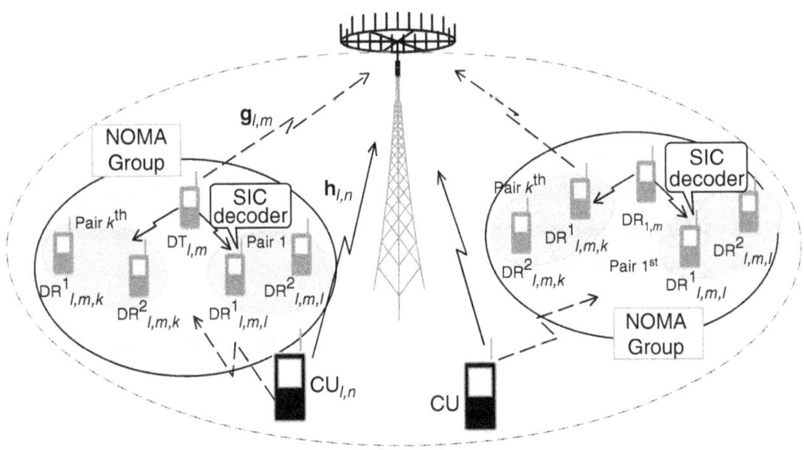

Figure 5.1 The proposed system model [42].

- *Downlink channels between cellular users, D2D transmitter devices, and D2D users:* $\{g^{(1)}_{l,m,k}, g^{(2)}_{l,m,k}\} \in \mathbb{C}$ is the channel between the D2D transmitter device in group m and the two users of pair k in group m operating in subchannel l. Additionally, $h^{(1)}_{l,n,m,k} \in \mathbb{C}$ and $h^{(2)}_{l,n,m,k} \in \mathbb{C}$ denotes the interfering channels between cellular user n utilizing subchannel l and two users of pair k in group m, respectively.

5.2.2 Capacity Analysis of D2D and Cellular Networks

During the uplink transmission of cellular users, $x_{l,n}$ denotes the data sent by cellular user n that is occupying subchannel l with $\mathbb{E}\{|x_{l,n}|^2\} = 1$. This data symbol is allocated a power level $P^{C,l}_n$, which satisfies

$$0 \leq P^C_{l,n} \leq P^C_{\max,l,n} \tag{5.1}$$

Meanwhile, in group m, the D2D transmitter is sending signals to its pairs. In particular, the data symbols $t^{(1)}_{m,k}$ and $t^{(2)}_{m,k}$ will be transmitted to D2D user 1 and user 2 of pair k in group m, respectively. The transmitted power levels to these data symbols are denoted as $P^{D,(1)}_{m,k}$ and $P^{D,(2)}_{m,k}$, which satisfy

$$0 \leq P^{D,(1)}_{m,k} \leq P^{D,(1)}_{\max,m,k} \tag{5.2}$$

$$0 \leq P^{D,(2)}_{m,k} \leq P^{D,(2)}_{\max,m,k} \tag{5.3}$$

where $P^{D,(1)}_{\max,m,k}$ and $P^{D,(2)}_{\max,m,k}$ represent the maximum power that the D2D transmitter device can use to send the data. From this, we also calculate the achievable rates of cellular users and D2D users.

5.2.2.1 Uplink Cellular Networks Transmission

The received signal at the subchannel l^{th} of BS is denoted as $\mathbf{y}^C_l \in \mathbb{C}^N$ and given by:

$$\mathbf{y}_l = \sum_{n=1}^{N_l} \sqrt{P^C_{l,n}} \mathbf{h}_{l,n} x_{l,n} + \sum_{m=1}^{M} \sum_{k=1}^{K_m} \eta_{l,m} \mathbf{g}_m \left(\sqrt{P^{D,(1)}_{m,k}} t^{(1)}_{m,k} + \sqrt{P^{D,(2)}_{m,k}} t^{(2)}_{m,k} \right) + \zeta_l \tag{5.4}$$

where $x_{l,n}$ represents the uplink data transmitted by CU n which uses the lth subchannel with $\mathbb{E}\{|x_{l,n}|^2\} = 1$. The data represented by the symbols $\{t^{(1)}_{m,k}, t^{(2)}_{m,k}\}$ are transmitted from the mth D2D device transmitter to the two users in group m. ζ_l is the additive white Gaussian noise (AWGN) at the BS on subchannel l, which distributes as $\zeta_l \sim \mathcal{CN}(\mathbf{0}, \sigma^2_{\text{UL}} \mathbf{I}_N)$. Here, $\eta_{l,m}$ is defined as:

$$\eta_{l,m} = \begin{cases} 1, & \text{if group } m \text{ uses subchannel } l \\ 0, & \text{Otherwise} \end{cases} \tag{5.5}$$

To decode the desired signal from cellular user n using subchannel l, we assume that the cellular BS utilizes a detection vector $\mathbf{w}_{l,n} \in \mathbb{C}^N$ as:

$$\begin{aligned} \hat{x}_{l,n} = \mathbf{w}^H_{l,n} \mathbf{y}_l = & \underbrace{\sqrt{P^C_{l,n}} \mathbf{w}^H_{l,n} \mathbf{h}_{l,n} x_{l,n}}_{\text{Desired signal}} + \underbrace{\sum_{n'=1, n' \neq n}^{N_l} \sqrt{P^C_{l,n'}} \mathbf{w}^H_{l,n} \mathbf{h}_{l,n'} x_{l,n'}}_{\text{Cellular interference}} \\ & + \underbrace{\sum_{m=1}^{M} \sum_{k=1}^{K_m} \eta_{l,m} \mathbf{w}^H_{l,n} \mathbf{g}_m \left(\sqrt{P^{D,1}_{m,k}} t^{(1)}_{m,k} + \sqrt{P^{D,(2)}_{m,k}} t^{(2)}_{m,k} \right)}_{\text{D2D interference}} + \underbrace{\mathbf{w}^H_{l,n} \zeta_l}_{\text{Noise}} \end{aligned} \tag{5.6}$$

In (5.6), the first term represents the desired signal and the second term denotes mutual interference from other cellular users. The last terms represent mutual interference from D2D users and thermal noise. The achievable rate of this cellular user is therefore demonstrated in Lemma 5.1.

Lemma 5.1 *If BS assumes a detection vector $\mathbf{w}_{l,n}$ for user n who is using the l-th subchannel, then its achievable rate is computed as:*

$$R_{l,n}^C = \log_2\left(1 + \gamma_{l,n}^C\right) \quad \text{(b/s/Hz)} \tag{5.7}$$

where $\gamma_{l,n}^C$ denotes the SINR at the cellular which is expressed as in (5.8) *as:*

$$\frac{P_{l,n}^C |\mathbf{w}_{l,n}^H \mathbf{h}_{l,n}|^2}{\sum\limits_{\substack{n'=1,\\ n' \neq n}}^{N_l} P_{l,n'}^C |\mathbf{w}_{l,n}^H \mathbf{h}_{l,n'}|^2 + \sum\limits_{m=1}^{M} \sum\limits_{k=1}^{K_m} \eta_{l,m} |\mathbf{w}_{l,n}^H \mathbf{g}_m|^2 \left(P_{m,k}^{D,(1)} + P_{m,k}^{D,(2)}\right) + \|\mathbf{w}_{l,n}\|_2^2 \sigma_{UL}^2} \tag{5.8}$$

Here, we emphasize that the achievable rate in Lemma 5.1 is applicable to any detection method under arbitrary channel environments.

5.2.2.2 Downlink NOMA-D2D Transmission

For D2D communications in pair k in group m, the user with the smaller channel gain is called the weaker user. Without loss of generality, we assume in each pair that user 1 is the weaker and user 2 is the stronger, then

$$|g_{m,k}^{(1)}|^2 \leq |g_{m,k}^{(2)}|^2 \tag{5.9}$$

When paired D2D users decode their desired signals, we further assume that the perfect SIC only applies to the stronger user. This suggests that the weaker user will employ conventional decoding methods. We then model the received signal for each of the two D2D users using subchannel l. In particular, the received signal of user 1 is formulated as:

$$\begin{aligned} y_{m,k}^{D,(1)} = {} & \underbrace{g_{m,k}^{(1)} \sqrt{P_{m,k}^{D,(1)}} t_{m,k}^{(1)}}_{\text{Desired signal}} + \underbrace{g_{m,k}^{(1)} \sqrt{P_{m,k}^{D,(2)}} t_{m,k}^{(2)}}_{\text{Intra pair interference}} + \underbrace{\sum_{n=1}^{N_l} \sqrt{P_{l,n}^C} h_{l,n,m,k}^{(1)} x_{l,n}}_{\text{Cellular interference}} \\ & + \underbrace{\sum_{\substack{m'=1,\\ m' \neq m}}^{M} \sum_{k=1}^{K_{m'}} \eta_{l,m'} g_{m',k}^{(1)} \left(\sqrt{P_{m',k}^{D,(1)}} t_{m',k}^{(1)} + \sqrt{P_{m',k}^{D,2}} t_{m',k}^{(2)}\right)}_{\text{Inter pair interference}} + \underbrace{\zeta_{m,k}^{(1)}}_{\text{Noise}} \end{aligned} \tag{5.10}$$

where $\zeta_{m,k}^{(1)} \sim \mathcal{CN}(0, \sigma_{\text{DL}}^2)$. In (5.10), the first term denotes the desired signal and the second term refers to the intra-pair interference. The third and fourth terms represent interference other the cellular users and D2D transmitter devices that also use subchannel l. The last term represents thermal noise.

Based on (5.10), we formulate Lemma 5.2 to obtain the received SINR at the $\text{DR}_{m,k}^{(1)}$ as in (5.12).

Lemma 5.2 *At pair k in group m, if user 1 is the weaker user, then by conventional decoding method, the achievable rate of this user is given by:*

$$R_{m,k}^{D,(1)} = \log_2\left(1 + \gamma_{m,k}^{D,(1)}\right) \quad \text{(b/s/Hz)} \tag{5.11}$$

where the SINR value for this user is denoted as: $\gamma_{m,k}^{D,(1)}$ *and shown in* (5.12) *as*

$$\frac{P_{m,k}^{D,(1)}|g_{m,k}^{(1)}|^2}{P_{m,k}^{D,(2)}|g_{m,k}^{(1)}|^2 + \sum_{n=1}^{N_l} P_{l,n}^C |h_{l,n,m,k}^{(1)}|^2 + \sum_{\substack{m'=1,\\ m'\neq m}}^{M} \sum_{k=1}^{K_{m'}} \eta_{l,m'}|g_{m',k}^{(1)}|^2 (P_{m',k}^{D,(1)} + P_{m',k}^{D,(2)}) + \sigma_{DL}^2} \tag{5.12}$$

In (5.12), the numerator indicates the importance of power control and how much signal is degraded over the propagation environment. The first term of the denominator denotes the power of intra-pair interference, while the second term represents the total power of all cellular users that share the same subchannel l. The third term represents mutual interference from other D2D groups and the last term represents thermal noise.

We now formulate the received signal at user 2 of pair k in group m as:

$$\begin{aligned} y_{m,k}^{D,(2)} = &\underbrace{g_{m,k}^{(2)}\sqrt{P_{m,k}^{D,(2)}}t_{m,k}^{(2)}}_{\text{Desired signal}} + \underbrace{g_{m,k}^{(2)}\sqrt{P_{m,k}^{D,(1)}}t_{m,k}^{(1)}}_{\text{Intra pair interference}} + \underbrace{\sum_{n=1}^{N_l}\sqrt{P_{l,n}^C}h_{l,n,m,k}^{(2)}x_{l,n}}_{\text{Cellular interference}} \\ &+ \underbrace{\sum_{\substack{m'=1,\\ m'\neq m}}^{M}\sum_{k=1}^{K_{m'}} \eta_{l,m'}g_{m',k}^{(2)}\left(\sqrt{P_{m',k}^{D,(1)}}t_{m',k}^{(1)} + \sqrt{P_{m',k}^{D,(2)}}t_{m',k}^{(2)}\right)}_{\text{Inter pair interference}} + \underbrace{\zeta_{m,k}^{(2)}}_{\text{Noise}} \end{aligned} \tag{5.13}$$

In contrast to user 1, user 2 utilizes the perfect SIC to mitigate intra-pair interference and therefore the achievable rate for user 2 is formulated as in Lemma 5.3.

Lemma 5.3 *By applying the perfect SIC, the achievable rate of user 2 of pair k in group m is*

$$R_{m,k}^{D,(2)} = \log_2\left(1 + \gamma_{m,k}^{D,(2)}\right) \quad \text{(b/s/Hz)} \tag{5.14}$$

where the SINR value of this user, denoted as $\gamma_{m,k}^{D,(2)}$, *is shown in* (5.15) *as*

$$\frac{P_{m,k}^{D,(2)}|g_{m,k}^{(2)}|^2}{\sum_{n=1}^{N_l} P_{l,n}^C |h_{l,n,m,k}^{(2)}|^2 + \sum_{\substack{m'=1,\\ m'\neq m}}^{M} \sum_{k=1}^{K_{m'}} \eta_{l,m'}|g_{m',k}^{(2)}|^2 (P_{m',k}^{D,(1)} + P_{m',k}^{D,(2)}) + \sigma_{DL}^2} \tag{5.15}$$

Lemma 5.3 demonstrates that the intra-pair interference from user 1 is completely mitigated using the perfect SIC. However, both lemmas confirm the existence of D2D inter-pair interference, which is effectively reduced by subchannel assignment.

5.3 Joint Subchannel Assignment and Power Control for D2D Communication

In this section, we outline the total transmit power consumption of the D2D users subject to the subchannel assignment and power control constraints.

The total transmit power minimization problem of all D2D transmitter devices and cellular users in the network is formulated as:

$$\begin{aligned}
\underset{\substack{\{\eta_{l,m}\in\{0,1\}\},\\ \{P^{D,(1)}_{m,k}\geq 0\},\\ \{P^{D,(2)}_{m,k}\geq 0\},\\ \{P^{C}_{l,n}\geq 0\}}}{\text{minimize}} \quad & \sum_{m=1}^{M}\sum_{k=1}^{K_m}\left(P^{D,(1)}_{m,k}+P^{D,(2)}_{m,k}\right)+\sum_{l=1}^{L}\sum_{n=1}^{N_l}P^{C}_{l,n}\\
\text{Subject to} \quad & R^{C}_{l,n}\geq R^{C,\text{thr}}_{l,n}, \quad \forall l,n\\
& R^{D,(1)}_{m,k}\geq R^{D,(1),\text{thr}}_{m,k}, \quad \forall m,k\\
& R^{D,(2)}_{m,k}\geq R^{D,(2),\text{thr}}_{m,k}, \quad \forall m,k\\
& P^{D,(1)}_{m,k}\leq P^{D,(1)}_{\max,m,k}, \quad \forall m,k\\
& P^{D,(2)}_{m,k}\leq P^{D,(2)}_{\max,m,k}, \quad \forall m,k\\
& P^{C}_{l,n}\leq P^{C}_{\max,l,n}, \quad \forall l,n
\end{aligned} \tag{5.16}$$

where $R^{C,\text{thr}}_{l,n}$, $R^{D,(1),\text{thr}}_{m,k}$, and $R^{D,(2),\text{thr}}_{m,k}$ represent the quality of service requirements from the cellular and D2D users which are measured in (b/s/Hz). By setting

$$\gamma^{C,\text{thr}}_{l,n}=2^{R^{C,\text{thr}}_{l,n}}-1 \tag{5.17}$$

$$\gamma^{D,(1),\text{thr}}_{m,k}=2^{R^{D,(1),\text{thr}}_{m,k}}-1 \tag{5.18}$$

$$\gamma^{D,(2),\text{thr}}_{m,k}=2^{R^{D,(2),\text{thr}}_{m,k}}-1 \tag{5.19}$$

we can then convert the optimization problem (5.16) from the quality of service constraints to the SINR constraints as:

$$\begin{aligned}
\underset{\substack{\{\eta_{l,m}\in\{0,1\}\},\\ \{P^{D,(1)}_{m,k}\geq 0\},\\ \{P^{D,(2)}_{m,k}\geq 0\},\\ \{P^{C}_{l,n}\geq 0\}}}{\text{minimize}} \quad & \sum_{m=1}^{M}\sum_{k=1}^{K_m}\left(P^{D,(1)}_{m,k}+P^{D,(2)}_{m,k}\right)+\sum_{l=1}^{L}\sum_{n=1}^{N_l}P^{C}_{l,n}\\
\text{Subject to} \quad & \gamma^{C}_{l,n}\geq \gamma^{C,\text{thr}}_{l,n}, \quad \forall l,n\\
& \gamma^{D,(1)}_{m,k}\geq \gamma^{D,(1),\text{thr}}_{m,k}, \quad \forall m,k\\
& \gamma^{D,(2)}_{m,k}\geq \gamma^{D,(2),\text{thr}}_{m,k}, \quad \forall m,k\\
& P^{D,(1)}_{m,k}\leq P^{D,(1)}_{\max,m,k}, \quad \forall m,k\\
& P^{D,(2)}_{m,k}\leq P^{D,(2)}_{\max,m,k}, \quad \forall m,k\\
& P^{C}_{l,n}\leq P^{C}_{\max,l,n}, \quad \forall l,n
\end{aligned} \tag{5.20}$$

We observe that the optimization problem (5.20) is a mixed integer program and therefore a combinatorial problem, which is inherently in extremely high computational complexity to obtain the global solution.[1] Moreover, we have so far

1 One way to get a solution with given ϵ-accuracy to problem (5.20) is utilizing the relaxation of binary optimization variables as what the outer approximation algorithm (OAA) does. However,

assumed that the D2D pairs are available. We now propose a pairing method that aims at minimizing the total transmit power of the D2D transmitter devices.

5.3.1 Subchannel Assignment Scheme

Under an assumption that all cellular users have already known their subchannels, this subsection proposes a heuristic algorithm to reuse subchannels in each D2D group. The proposed subchannel assignment algorithm should satisfy the following criteria:

- The number of groups on a channel is as minimal as possible in order to minimize the co-channel interference between different groups as well as to reduce the complexity of the general resource allocation.
- Device groups will be assigned into subchannels so that the influence of cellular devices is minimized. This arrangement takes advantage of both large- and small-scale fading factors to minimize interference from CU to D2D groups.

Algorithm 5.1 Subchannel assignment for D2D users based on diversity gain

1: **Input**: Initial number of subchannels L, number of groups M
2: For each group m, compute the set S_m as in (5.21)
3: **if** $M \leq L$ **then**
4: {On each frequency channel l there exists only one D2D group}
5: Find the free subchannel in S_m having the minimum $\sum_{n=1}^{N_l}\sum_{k=1}^{K_m}\left(|h^{(1)}_{l,n,m,k}| + |h^{(2)}_{l,n,m,k}|\right)$.
6: **else**
7: {$M > L$: Exist more than one reusable D2D group on a subchannel}
8: **If** M Mod $L = 0$ **then** each subchannel having M/L groups: Repeat Steps 2, 3, 4, 5 to assign subchannels to subsets of L groups. **Otherwise,** $L - (M - L\lfloor M/L \rfloor)$ subchannels have $\lfloor M/L \rfloor$ D2D groups, while $M - L\lfloor M/L \rfloor$ subchannel has $\lfloor M/L \rfloor + 1$ D2D groups.
9: **end if**
10: **Output**: Subchannel assignment information for all D2D users.

Source: Yoon et al. [42].

Algorithm 5.1 expresses the proposed subchannel assignment. In particular, for each D2D group m, we compute the set S_m, which contains the total channel gain from the cellular users in a subchannel and all D2D groups as demonstrated in (5.21).

$$S_m = \left\{ \sum_{n=1}^{N_1}\sum_{k=1}^{K_m}\left(|h^{(1)}_{1,n,m,k}| + |h^{(2)}_{1,l,m,k}|\right), \ldots, \sum_{n=1}^{N_L}\sum_{k=1}^{K_m}\left(|h^{(1)}_{L,n,m,k}| + |h^{(2)}_{L,n,m,k}|\right) \right\} \tag{5.21}$$

the relaxed problem is only convex after the hidden convex structure is exploited with high computational complexity, while the solution is only a local optimum to (5.20). In this paper, we would like to propose a low complexity algorithm to evaluate the performance of large-scale networks. The comparison of our proposed method and OAA is left for future work.

The relationship of total subchannels L and the number of D2D groups M can then be divided into two cases. The first case is when $M \le L$, in which there are sufficient frequency resources for each D2D group occupying one subchannel. For D2D group m, the subchannel is selected to ensure the minimum diversity gain, i.e. the minimal value of $\sum_{n=1}^{N_l}\sum_{k=1}^{K_m}\left(|h_{l,n,m,k}^{(1)}|+|h_{l,n,m,k}^{(2)}|\right)$ as shown in Step 5. This assignment satisfies the two previously outlined goals. In the case $M > L$, there is more than one D2D group sharing the same subchannel. If M represents an integer times of L, then each subchannel will be utilized by M/L D2D groups. Otherwise, the division of M/L is not an integer, each subchannel is assigned $\lfloor M/L \rfloor$ D2D groups. After that, each of the remaining $M - L\lfloor M/L \rfloor$ D2D groups will be allocated to a subchannel that satisfies the minimal interference effected by cellular users on D2D users.

5.3.2 Power Control Scheme

For a given set of $\{\eta_{l,m}\}$, the total transmit power minimization problem (5.20) becomes

$$
\begin{aligned}
\underset{\substack{\{P_{m,k}^{D,(1)},\\ P_{m,k}^{D,(2)},\\ P_{l,n}^{C}\ge 0\}}}{\text{minimize}} \quad & \sum_{m=1}^{M}\sum_{k=1}^{K_m}\left(P_{m,k}^{D,(1)}+P_{m,k}^{D,(2)}\right)+\sum_{l=1}^{L}\sum_{n=1}^{N_l}P_{l,n}^{C} \\
\text{Subject to} \quad & \gamma_{l,n}^{C}\ge\gamma_{l,n}^{C,\text{thr}}, \quad \forall l,n \\
& \gamma_{m,k}^{D,(1)}\ge\gamma_{m,k}^{D,(1),\text{thr}}, \quad \forall m,k \\
& \gamma_{m,k}^{D,(2)}\ge\gamma_{m,k}^{D,(2),\text{thr}}, \quad \forall m,k \\
& P_{m,k}^{D,(1)}\le P_{\max,m,k}^{D,(1)}, \quad \forall m,k \\
& P_{m,k}^{D,(2)}\le P_{\max,m,k}^{D,(2)}, \quad \forall m,k \\
& P_{l,n}^{C}\le P_{\max,l,n}^{C}, \quad \forall l,n
\end{aligned}
\tag{5.22}
$$

Equation (5.22) is a convex program as shown in Theorem 5.1.

Theorem 5.1 *The optimization problem* (5.22) *is a linear program. Consequently, if the global solution exists, it can be obtained in polynomial time by utilizing any general interior point toolbox.*

Proof: The main proof is to show that (5.22) is a linear program. In fact, the objective function is linear. The power budget constraints are also linear, making the feasible set convex. Moreover, the SINR constraints of cellular users can be rewritten as:

$$
\begin{aligned}
&\gamma_{l,n}^{C,\text{thr}}\sum_{n'=1}^{N_l}P_{l,n'}^{C}|\mathbf{w}_{l,n}^{H}\mathbf{h}_{l,n'}|^2+\gamma_{l,n}^{C,\text{thr}}\sum_{m=1}^{M}\sum_{k=1}^{K_m}\eta_{l,m}|\mathbf{w}_{l,n}^{H}\mathbf{g}_m|^2\left(P_{m,k}^{D,(1)}+P_{m,k}^{D,(2)}\right)\\
&\quad+\gamma_{l,n}^{C,\text{thr}}\|\mathbf{w}_{l,n}\|_2^2\sigma_{\text{UL}}^2-\left(\gamma_{l,n}^{C,\text{thr}}+1\right)P_{l,n}^{C}|\mathbf{w}_{l,n}^{H}\mathbf{h}_{l,n}|^2\le 0
\end{aligned}
\tag{5.23}
$$

which is an affine constraint. The SINR constraints of D2D users are also affine by using similar manipulations. Therefore, aligned with the standard form [43, 44], the optimization problem (5.22) is a linear program. ■

Theorem 5.1 manifests the usefulness of a general purpose toolbox such as CVX to find the global solution to (5.22) [45]. However, with $2\sum_{m=1}^{M} K_m + \sum_{l=1}^{L} N_l$ optimization variables and $2\sum_{l=1}^{L} N_l + 4\sum_{m=1}^{M} K_m$ constraints, this toolbox using the interior point method requires computational complexity in the order of:

$$\mathcal{O}\left(\max\left\{F_1, F_2\right\}\right) \tag{5.24}$$

where F_1 and F_2 are defined as

$$F_1 = \left(2\sum_{m=1}^{M} K_m + \sum_{l=1}^{L} N_l\right)^3 \tag{5.25}$$

$$F_2 = \left(2\sum_{m=1}^{M} K_m + \sum_{l=1}^{L} N_l\right)^2 \left(2\sum_{l=1}^{L} N_l + 4\sum_{m=1}^{M} K_m\right) \tag{5.26}$$

which still increases exponentially with the network size. Therefore, in this paper, we investigate a closed-form solution to (5.22) based on the observation that if the optimal solution exists, then all SINR constraints will be equal. Specifically, the $\sum_{l=1}^{L} N_l$ SINR constraints of cellular users becomes equal as shown below:

$$\begin{aligned} &\frac{P_{l,n}^{C} |\mathbf{w}_{l,n}^{H} \mathbf{h}_{l,n}|^2}{\gamma_{l,n}^{C,\text{thr}}} - \sum_{n'=1, n' \neq n}^{N_l} P_{l,n'}^{C} |\mathbf{w}_{l,n}^{H} \mathbf{h}_{l,n'}|^2 - \sum_{m=1}^{M} \sum_{k=1}^{K_m} \eta_{l,m} \\ &\times |\mathbf{w}_{l,n}^{H} \mathbf{g}_m|^2 \left(P_{m,k}^{D,(1)} + P_{m,k}^{D,(2)}\right) - \|\mathbf{w}_{l,n}\|_2^2 \sigma_{\text{UL}}^2 = 0, \quad \forall l, n \end{aligned} \tag{5.27}$$

Similarly, the $\sum_{m=1}^{M} K_m$ SINR constraints of weaker D2D users becomes

$$\begin{aligned} &\frac{P_{m,k}^{D,(1)} |g_{m,k}^{(1)}|^2}{\gamma_{m,k}^{D,(1)}} - P_{m,k}^{D,(2)} |g_{m,k}^{(1)}|^2 - \sum_{\substack{m'=1 \\ m' \neq m}}^{M} \sum_{k=1}^{K_{m'}} \eta_{l,m'} |g_{m',k}^{(1)}|^2 \left(P_{m',k}^{D,(1)} + P_{m',k}^{D,(2)}\right) \\ &- \sum_{n=1}^{N_l} P_{l,n}^{C} |h_{l,n,m,k}^{(1)}|^2 - \sigma_{\text{DL}}^2 = 0, \quad \forall m, k \end{aligned} \tag{5.28}$$

In the optimal solution, the $\sum_{m=1}^{M} K_m$ SINR constraint of stronger D2D users is equivalent to:

$$\begin{aligned} &\frac{P_{m,k}^{D,(2)} |g_{m,k}^{(2)}|^2}{\gamma_{m,k}^{D,(2)}} - \sum_{\substack{m'=1 \\ m' \neq m}}^{M} \sum_{k=1}^{K_{m'}} \eta_{l,m'} |g_{m',k}^{(2)}|^2 \left(P_{m',k}^{D,(1)} + P_{m',k}^{D,(2)}\right) \\ &- \sum_{n=1}^{N_l} P_{l,n}^{C} |h_{l,n,m,k}^{(2)}|^2 - \sigma_{\text{DL}}^2 = 0, \quad \forall, m, k \end{aligned} \tag{5.29}$$

We emphasize that solving a system of linear equations in (5.27), (5.28), and (5.29) are able to be reduced to a matrix inversion computation. By utilizing the Cholesky decomposition method [46], the computational complexity is reduced to:

$$\mathcal{O}\left(\left(2\sum_{m=1}^{M} K_m + \sum_{l=1}^{L} N_l\right)^2\right) \tag{5.30}$$

which is less complex than the interior point method.

5.4 Optimization of D2D Device Pairing

By assuming that all cellular users have been allocated to subchannels, we now propose a method to pair the D2D users. The main goal is to minimize the mutual interference between cellular users and D2D users using the system diversity guidelines as shown in Lemma 5.4.

Lemma 5.4 *In each D2D group m, we first arrange the channel gains of all* $2K_m$ *users before pairing as:*

$$\begin{aligned} &|f_{m,1}| \leq |f_{m,2}| \leq \cdots \leq |f_{m,K_m-1}| \leq |f_{m,K_m}| \\ &\leq |f_{m,K_m+1}| \leq \cdots \leq |f_{m,2K_m}| \end{aligned} \tag{5.31}$$

then we split these users into two groups of the user indices:

$$S_1 = \{1,2,\ldots,K_m\}, \quad S_2 = \{K_m, K_{m+1}, \ldots, K_{2m}\} \tag{5.32}$$

The paring method $\{1, K_m\}$, $\{2, K_{m+1}\}, \ldots,$ $\{K_m, K_{2m}\}$ *minimizes the total transmit power of all D2D transmitter devices.*

Proof: We first rewrite the SINR formula of D2D user 2 of pair k in group m as follows:

$$\gamma_{m,k}^{D,(2)} = \frac{P_{m,k}^{D,(2)} |g_{m,k}^{(2)}|^2}{I_{m,k}^{C,(2)} + I_{m,k}^{D,(2)} + \sigma_{\text{DL}}^2} \tag{5.33}$$

where the mutual interference $I_{m,k}^{C,(2)}$ and $I_{m,k}^{D,(2)}$ are respectively defined as:

$$I_{m,k}^{C,(2)} = \sum_{n=1}^{N_l} P_{l,n}^C |h_{l,n,m,k}^{(1)}|^2, \quad I_{m,k}^{D,(2)} = \sum_{\substack{m'=1 \\ m' \neq m}}^{M} \sum_{k=1}^{K_{m'}} \eta_{l,m'} |g_{m',k}^{(2)}|^2 (P_{m',k}^{D,(1)} + P_{m',k}^{D,(2)}) \tag{5.34}$$

From the SINR constraint in the optimization problem (5.20), the data power level for this user the satisfies

$$P_{m,k}^{D,(2)} \geq \gamma_{m,k}^{D,(2)} \frac{I_{m,k}^{C,(2)} + I_{m,k}^{D,(2)} + \sigma_{\text{DL}}^2}{|g_{m,k}^{(2)}|^2} \tag{5.35}$$

Similarly, the SINR formula of D2D user 1 of pair k in group m as the following:

$$\gamma_{m,k}^{D,(1)} = \frac{P_{m,k}^{D,(1)} |g_{m,k}^{(1)}|^2}{P_{m,k}^{D,(2)} |g_{m,k}^{(1)}|^2 + I_{m,k}^{C,(2)} + I_{m,k}^{D,(1)} + \sigma_{\text{DL}}^2} \tag{5.36}$$

where $I_{m,k}^{C,1}$ and $I_{m,k}^{D,(1)}$ are respectively defined as:

$$I_{m,k}^{C,(1)} = \sum_{n=1}^{N_l} P_{l,n}^C |h_{l,n,m,k}^{(1)}|^2 \tag{5.37}$$

$$I_{m,k}^{D,(1)} = \sum_{\substack{m'=1 \\ m' \neq m}}^{M} \sum_{k=1}^{K_{m'}} \eta_{l,m'} |g_{m',k}^{(1)}|^2 (P_{m',k}^{D,(1)} + P_{m',k}^{D,(2)}) \tag{5.38}$$

Therefore, due to its SINR constraint in the optimization problem (5.20), the data power level of this user must equally satisfy the following:

$$P^{D,(1)}_{m,k} \geq \gamma^{D,(1)}_{m,k} \frac{P^{D,(2)}_{m,k}|g^{(1)}_{m,k}|^2 + I^{C,(1)}_{m,k} + I^{D,(1)}_{m,k} + \sigma^2_{\text{DL}}}{|g^{(1)}_{m,k}|^2} \tag{5.39}$$

$$\geq \gamma^{D,(1)}_{m,k} \frac{\gamma^{D,(2)}_{m,k} \frac{I^{C,(2)}_{m,k}+I^{D,(2)}_{m,k}+\sigma^2_{\text{DL}}}{|g^{(2)}_{m,k}|^2} |g^{(1)}_{m,k}|^2 + I^{C,(1)}_{m,k} + I^{D,(1)}_{m,k} + \sigma^2_{\text{DL}}}{|g^{(1)}_{m,k}|^2} \tag{5.40}$$

$$= \gamma^{D,(1)}_{m,k}\gamma^{D,(2)}_{m,k} \frac{I^{C,(2)}_{m,k} + I^{D,(2)}_{m,k} + \sigma^2_{\text{DL}}}{|g^{(2)}_{m,k}|^2} + \gamma^{D,(1)}_{m,k} \frac{I^{C,(1)}_{m,k} + I^{D,(1)}_{m,k} + \sigma^2_{\text{DL}}}{|g^{(1)}_{m,k}|^2} \tag{5.41}$$

where (5.40) is attained by using (5.35) into (5.39). Combining (5.35) and (5.40), the total transmit power of all D2D users is bounded by:

$$\begin{aligned}
&\sum_{m=1}^{M}\sum_{k=1}^{K_m}\left(P^{D,(1)}_{m,k} + P^{D,(2)}_{m,k}\right) \\
&\geq \sum_{m=1}^{M}\sum_{k=1}^{K_m}\gamma^{D,(1)}_{m,k}\gamma^{D,(2)}_{m,k} \frac{I^{C,(2)}_{m,k} + I^{D,(2)}_{m,k} + \sigma^2_{\text{DL}}}{|g^{(2)}_{m,k}|^2} + \sum_{m=1}^{M}\sum_{k=1}^{K_m}\gamma^{D,(1)}_{m,k} \frac{I^{C,(1)}_{m,k} + I^{D,(1)}_{m,k} + \sigma^2_{\text{DL}}}{|g^{(1)}_{m,k}|^2} \\
&+ \sum_{m=1}^{M}\sum_{k=1}^{K_m}\gamma^{D,(2)}_{m,k} \frac{I^{C,(2)}_{m,k} + I^{D,(2)}_{m,k} + \sigma^2_{\text{DL}}}{|g^{(2)}_{m,k}|^2} \qquad (5.42)\\
&\geq \alpha\left(\sum_{m=1}^{M}\sum_{k=1}^{K_m}\frac{1}{|g^{(2)}_{m,k}|^2} + \sum_{m=1}^{M}\sum_{k=1}^{K_m}\left(\frac{1}{|g^{(1)}_{m,k}|^2} + \frac{1}{|g^{(2)}_{m,k}|^2}\right)\right) \qquad (5.43)
\end{aligned}$$

where (5.42) is obtained by using the lower bounds of $P_{m,k,1}$ and $P_{m,k,2}$ in (5.41) and (5.35), respectively. The constant value α is given by:

$$\begin{aligned}
\alpha = \min_{l,k} \Big\{&\gamma^{D,(1)}_{m,k}\gamma^{D,(2)}_{m,k}\left(I^{C,(2)}_{m,k} + I^{D,(2)}_{m,k} + \sigma^2_{\text{DL}}\right),\ \gamma^{D,(1)}_{m,k}\left(I^{C,(1)}_{m,k} + I^{D,(1)}_{m,k} + \sigma^2_{\text{DL}}\right),\\
&\gamma^{D,(2)}_{m,k}\left(I^{C,(2)}_{m,k} + I^{D,(2)}_{m,k} + \sigma^2_{\text{DL}}\right)\Big\} \qquad (5.44)
\end{aligned}$$

This takes into consideration that in (5.43), the summation of the last two terms is independent of pairing method, meaning that

$$\sum_{m=1}^{M}\sum_{k=1}^{K_m}\left(\frac{1}{|g^{(1)}_{m,k}|^2} + \frac{1}{|g^{(2)}_{m,k}|^2}\right) = \text{constant} \tag{5.45}$$

As a consequence, the result implies that the total power consumption of all D2D users is a function of the first term in (5.43), which we can prove using the proposed pairing method. Therefore, we complete the proof. ■

5.5 Results and Discussion

In this section, the theoretical analysis and proposed methods from the previous sections are qualified and discussed using numerical results. In particular, we are

Table 5.1 Simulated network parameters.

Cellular radius	**250 m**
Base station's height	6 m
Beamforming technology	MRC
D2D group radius	10 m
Device and cellular user's height	1.5 m
Operating Frequency	1.9 GHz
Maximum transmit power of device and cellular user	23 dBm
Noise power	−96 dBm
Number of subchannel	5
Quality of service	1 b/s/Hz

Source: Yoon et al. [42].

interested in dividing pairs in groups and simulating algorithms using a common subset of subchannels. We consider the performance of the joint subchannel and pairing in four cases included diversity-based method (DBM) with proposed pairing, DBM with conventional pairing, random channel assignment with proposed pairing, and random channel assignment with conventional pairing.

The general parameters of the network are given in Table 5.1, while a number of specific parameters used for particular simulation are described separately for each case. Due to the simple and scalable structure, the BS uses maximum ratio combining (MRC) to decode transmitted signals from cellular users.

5.5.1 Channel Model

In the network, all propagation channels follow uncorrelated Rayleigh distribution patterns and are formulated as:

$$\mathbf{h}_{l,n} \sim \mathcal{CN}\left(\mathbf{0}, \beta_{l,n}\mathbf{I}_N\right), \quad \forall n, l, \quad \mathbf{g}_m \sim \mathcal{CN}\left(\mathbf{0}, \beta_m\mathbf{I}_N\right), \quad \forall m \tag{5.46}$$

$$h^{(i)}_{l,n,m,k} \sim \mathcal{CN}\left(0, \beta^{(i)}_{l,n,m,k}\right), \quad \forall l, n, m, k, i, \quad g^{(i)}_{m,k} \sim \mathcal{CN}\left(0, \beta^{(i)}_{m,k}\right), \quad \forall l, m, i \tag{5.47}$$

where $\beta_{l,n}$, β_m, $\beta^{(i)}_{l,n,m,k}$, and $\beta^{(i)}_{m,k}$ represent large-scale fading coefficients modeling the path loss and shadow fading. They are respectively defined as:

$$\beta_{l,n} = \xi_{l,n}10^{\frac{\sigma z_{l,n}}{10}}, \beta_m = \xi_m 10^{\frac{\sigma z_m}{10}}, \beta^{(i)}_{l,n,m,k} = \xi^{(i)}_{l,n,m,k}10^{\frac{\sigma z_{l,n,m,k}}{10}}, \beta^{(i)}_{m,k} = \xi^{(i)}_{m,k}10^{\frac{\sigma z_{m,k}}{10}} \tag{5.48}$$

where σ is the standard deviation of shadow fading, which is set to 7 dB in this paper. $z_{l,n}$, z_m, $z_{l,n,m,k}$, and $z_{m,k}$ are distributed as $\mathcal{CN}(0,1)$. For the path-loss coefficients $\xi_{l,n}$, ξ_m, $\xi^{(i)}_{l,n,m,k}$, and $\xi^{(i)}_{m,k}$, we use the dual-slope model which is given by:

$$\xi = \begin{cases} -47.85 - 20\log 10(d), & \text{if } 0 < d \leq d_c\ (m) \\ -47.85 - 20\log_{10}(d_c) - 40\log_{10}(d/d_c), & \text{otherwise} \end{cases} \tag{5.49}$$

where ξ (dBm) $\in \{\xi_{l,n}, \xi_m, \xi^{(i)}_{l,n,m,k}, \xi^{(i)}_{m,k}\}$ and d (measured in meters) represent the distance between transmitter and receiver, i.e. $d \in \{d_{l,n}, d_m, d^{(i)}_{l,n,m,k}, d^{(i)}_{m,k}\}$. For the BS to cellular or BS to NOMA-D2D devices communication, the threshold d_c is set to 152 m, while $d_c = 25$ m represents the threshold for NOMA-D2D communication or cellular to D2D devices.

5.5.2 Performance Evaluation

Figure 5.2 visualizes the total transmit power of D2D devices versus the number of pairs in each NOMA-D2D group. In a general trend, the total transmit power increases with the number of D2D devices in a group. In particular, the network using the proposed pairing consumes the total transmit power from 14 mW up to 58 mW when the number of pairs in each group increases from 2 to 10. Additionally, the conventional pairing with the same distribution of pairs requires a total transmit power from 16 to 66 mW. Without pairing, we observed that the total transmit power dramatically grows up approximately 1282 times when the number of pairs increases by fivefold (i.e. from 2 pairs to 10 pairs). In summary, the proposed pairing method consumes 5% to 10% power less than that of the conventional pairing. Therefore, this result demonstrates the effectiveness of pairing methods for D2D devices.

We demonstrate in Figure 5.3 that the number of accessed D2D pairs as a function of cellular users. This shows how many D2D pairs can access the network while maintaining service quality for each user. Because the time and frequency resources are shared, the number of accessed D2D pairs will decrease as the number of cellular users on the network increases. The proposed method is the slowest, defined by a reduction from 10 to 2 D2D pairs. Additionally, the conventional pairing method does not allow any D2D pair to access the network when 10 cellular users are active.

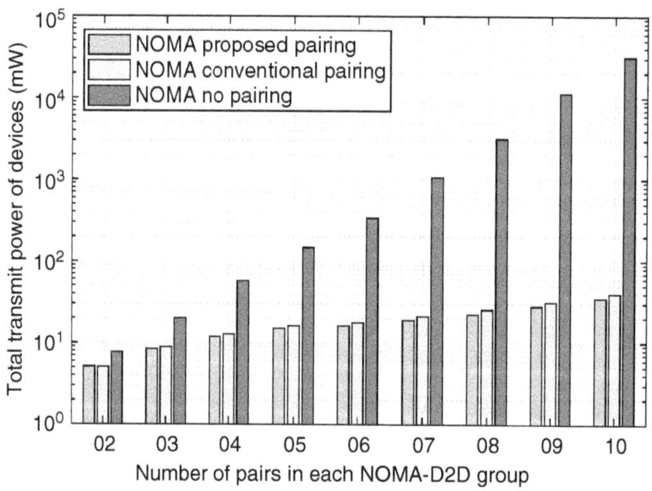

Figure 5.2 Total transmit power of D2D devices various number of pairs in each NOMA-D2D group [42].

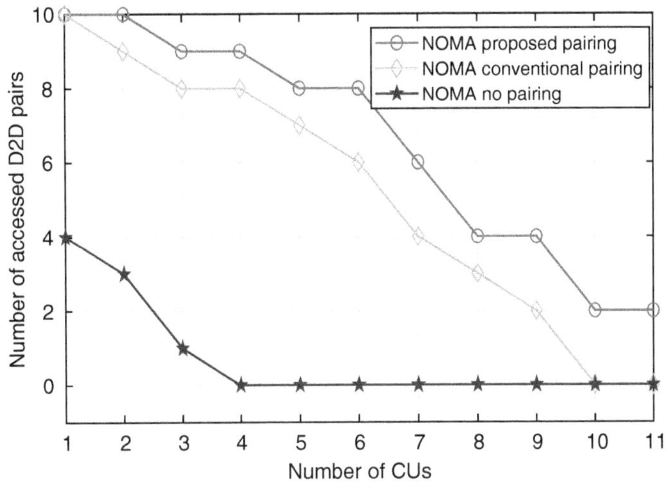

Figure 5.3 Number of accessed D2D pairs versus number of cellular users [42].

Specifically, in the absent of a defined pairing method, D2D devices cannot access the network if only four cellular users are available. This emphasizes that the two pairing methods detailed in this manuscript improve the performance compared to no pairing, and that the pairing proposed in this manuscript outperforms conventional pairing methods. Our proposed pairing method is more effective because it permits more D2D devices to join the network than other method.

Figure 5.4 shows the total transmit power of D2D devices with different numbers of cellular users. The total transmit power is directly proportional to the number of cellular users if the network only serves a subset, e.g. five cellular users without pairing. For each method, there is an upper bound of the number of cellular users that the transmit power of D2D devices saturates. Because new D2D users cannot

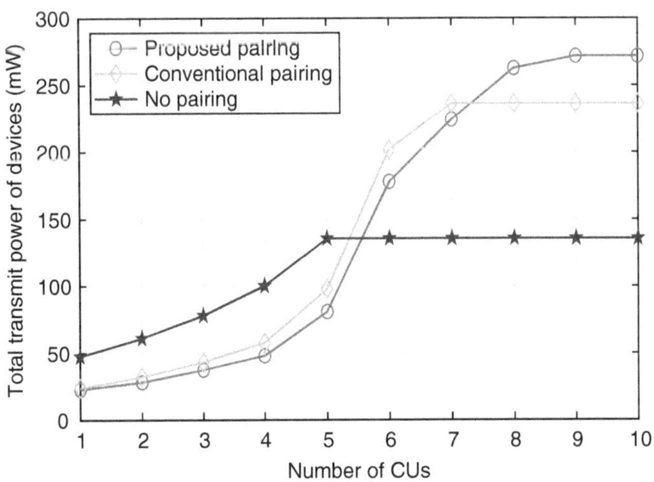

Figure 5.4 Total transmit power of D2D devices versus number of cellular users [42].

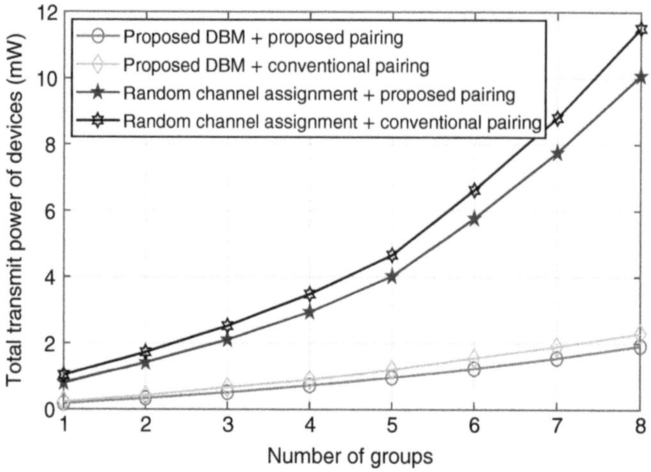

Figure 5.5 Total transmit power of D2D devices versus various number of D2D groups. The network is serving five cellular users, each of them activating in one separate subchannel [42].

enter the network, the limits for the conventional and proposed pairing methods are seven and nine users, respectively. This phenomenon comes from the fact that no new D2D users can enter the network. Whenever one tries to do so, the system falls in infeasible domain due to quality of service maintenance and due to limited power budget at every user as well. For a network with 10 cellular users, our system can serve more D2D users because our maximum transmit power for D2D users is 270 mW which is higher than the 240 mW available with the conventional pairing method.

Figure 5.5 illustrates the relationship of the total transmit power of D2D devices and the number of groups. This figure demonstrates the effectiveness of combining subchannel assignment with DBM, as in Algorithm 5.1. First, the transmit power increases with the number of groups. However, the combining DBM and our proposed pairing method requires the total transmit power to increase slowly from 0.2 to 1.9 mW when the number of groups increase from 1 to 8. This contrasts with DBM combined with conventional pairing, in which the power consumption increases from 0.25 to 2.4 mW. Without DBM, the network consumes the most power in all considered scenarios. Using only the proposed pairing method without DBM, D2D devices require 10.2 mW to sending data, while conventional pairing requires 11.5 mW. In summary, the proposed method of assignment reduces the transmit power up to five times in the tested scenarios compared to no pairing.

Figure 5.6 provides the accessed probability of NOMA-D2D system as the number of D2D groups increases. Our proposed pairing method allows all D2D devices to access the network as long as the number of groups does not exceed the number of available subchannels. If the number of D2D groups is greater than the number of subchannels, the combination of DBM and the proposed pairing slightly outperforms a combination of DBM and conventional pairing. This figure indicates the importance of using an advanced channel assignment because random channel

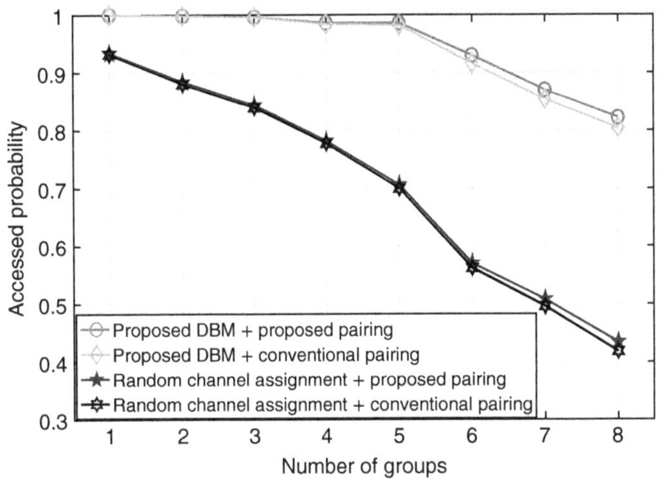

Figure 5.6 The accessed probability of NOMA-D2D system versus the number of D2D Group. The network has five subchannels and the service quality for every transmitters is 1 b/s/Hz [42].

assignment performs more poorly than our proposed DBM. The accessed probability by using DBM is better than random channel assignment by up to 39%.

5.6 Summary

We have investigated a network with coexistence of D2D devices and cellular users. Not only can the D2D pairs share the same resource with a subset of cellular users, but NOMA technology also enhances the SE. From the network topology, we have calculated the achievable rate for both D2D devices and cellular users. While considering both mutual interference as well as beamforming efficiency. We then formulated a total transmit power minimization problem that was subjected to massive constraints for all users available in the network. Even though this problem is NP-hard, we have divided it into subproblems that each of them can be efficiently solved with low complexity. Numerical results demonstrate the important roles of pairing methods and subchannel assignment in reducing the total transmit power of the network that serves both D2D devices and cellular users.

References

1 Doppler, K., Rinne, M., Wijting, C. et al. (2009). Device-to-device communication as an underlay to LTE-advanced networks. *IEEE Communications Magazine* 47 (12): 42–49.

2 Huynh, T., Onuma, T., Kuroda, K. et al. (2016). Joint downlink and uplink interference management for device to device communication underlaying cellular networks. *IEEE Access* 4: 4420–4430.

3 Hoa, N.T., Muzamane, H.A., and Van Duc, N. (2017). Geometric progression algorithm for adjacent band power allocation in OFDM based cognitive radio. *Proceedings of the 11th International Conference on Ubiquitous Information Management and Communication*, IMCOM '17, pp. 55:1–55:5. ACM.

4 Han, H.-D., Nguyen, D.-A., Nguyen, V.-D. et al. (2018). Pilot decontamination for multi-cell massive MIMO systems using asynchronous pilot design and data-aided channel estimation. *Physical Communication* 30: 76–85.

5 Mohammadi, M., Shi, X., Chalise, B.K. et al. (2017). Full-Duplex Non-Orthogonal Multiple Access for Modern Wireless Networks. *arXiv preprint arXiv:1710.11062.*

6 Tabassum, H., Ali, M.S., Hossain, E. et al. (2016). Non-Orthogonal Multiple Access (NOMA) in Cellular Uplink and Downlink: Challenges and Enabling Techniques. *arXiv preprint arXiv:1608.05783.*

7 Islam, S.M.R., Zeng, M., Dobre, O.A., and Kwak, K. (2018). Resource allocation for downlink NOMA systems: key techniques and open issues. *IEEE Wireless Communications* 25 (2): 40–47.

8 Ali, M.S., Tabassum, H., and Hossain, E. (2016). Dynamic user clustering and power allocation for uplink and downlink non-orthogonal multiple access (NOMA) systems. *IEEE Access* 4: 6325–6343.

9 Ali, S., Hossain, E., and Kim, D.I. (2017). Non-orthogonal multiple access (NOMA) for downlink multiuser MIMO systems: user clustering, beamforming, and power allocation. *IEEE Access* 5 (16705808): 565–577.

10 Zhao, J., Liu, Y., Chai, K.K. et al. (2016). NOMA-based D2D communications: towards 5G. *Global Communications Conference (GLOBECOM)*, Number 16655075, pp. 1–6.

11 Liu, Y., Qin, Z., Elkashlan, M. et al. (2017). Non-orthogonal multiple access for 5G and beyond. *Proceedings of the IEEE* 105 (12): 2347–2381.

12 Ding, Z., Lei, X., Karagiannidis, G.K. et al. (2017). A survey on non-orthogonal multiple access for 5G networks: research challenges and future trends. *IEEE Journal on Selected Areas in Communications* 35 (10): 2181–2195.

13 Islam, S.M.R., Avazov, N., Dobre, O.A., and Kwak, K. (2017). Power-domain non-orthogonal multiple access (NOMA) in 5G systems: potentials and challenges. *IEEE Communication Surveys and Tutorials* 19 (2): 721–742.

14 Ding, Z., Yang, Z., Fan, P., and Poor, H.V. (2014). On the performance of non-orthogonal multiple access in 5G systems with randomly deployed users. *IEEE Signal Processing Letters* 21 (12): 1501–1505.

15 Saito, Y., Benjebbour, A., Kishiyama, Y., and Nakamura, T. (2013). System-level performance evaluation of downlink non-orthogonal multiple access (NOMA). *24th International Symposium on Personal Indoor and Mobile Radio Communications (PIMRC)*, Number 14023675, pp. 611–615.

16 Saito, Y., Kishiyama, Y., Benjebbour, A. et al. (2013). Non-orthogonal multiple access (NOMA) for cellular future radio access. *IEEE Vehicular Technology Conference (VTC Spring)*, pp. 1–5.

17 Higuchi, K. and Benjebbour, A. (2015). Non-orthogonal multiple access (NOMA) with successive interference cancellation for future radio access. *IEICE Transactions on Communications* 98 (3): 403–414.

18 Benjebbour, A., Saito, K., Li, A. et al. (2015). Non-orthogonal multiple access (NOMA): concept, performance evaluation and experimental trials. *International Conference on Wireless Networks and Mobile Communications (WINCOM)*, pp. 1–6.

19 Parida, P. and Das, S.S. (2014). Power allocation in OFDM based NOMA systems: ADC programming approach. *GLOBECOM Workshops (GC Wkshps)*, pp. 1026–1031.

20 Chen, Z., Ding, Z., Dai, X., and Karagiannidis, G.K. (2016). On the application of quasi-degradation to MISO-NOMA downlink. *IEEE Transactions on Signal Processing* 64 (23): 6174–6189.

21 Zhao, J., Liu, Y., Chai, K.K. et al. (2017). Joint subchannel and power allocation for NOMA enhanced D2D communications. *IEEE Transactions on Communications* 65 (11): 5081–5094.

22 Pan, Y., Pan, C., Yang, Z., and Chen, M. (2018). Resource allocation for D2D communications underlaying a NOMA-based cellular network. *IEEE Wireless Communication Letter* 7: 130–133.

23 Elouafadi, R. and Benjillali, M. (2018). Cooperative NOMA-based D2D communications: a survey in the 5G/IoT context. *19th IEEE Mediterranean Electrotechnical Conference (MELECON)*, pp. 132–137.

24 Zhou, F., Wu, Y., Hu, R.Q. et al. (2018). Energy-efficient NOMA enabled heterogeneous cloud radio access networks. *IEEE Network* 32 (2): 152–160.

25 Kim, J.-B., Lee, I.-H., and Lee, J.H. (2018). Capacity scaling for D2D aided cooperative relaying systems using NOMA. *IEEE Wireless Communications Letters* 7 (1): 42–45.

26 Kang, J.-M. and Kim, I.-M. (2018). Optimal user grouping for downlink NOMA. *IEEE Wireless Communications Letters*. 7 (5): 724–727.

27 Sedaghat, M.A. and Müller, R.R. (2018). On user pairing in NOMA uplink. *IEEE Transactions on Wireless Communications* 17 (5): 3474–3486.

28 Wang, L., Tang, H., Wu, H., and Stüber, G.L. (2017). Resource allocation for D2D communications underlay in Rayleigh fading channels. *IEEE Transactions on Vehicular Technology* 66 (2): 1159–1170.

29 Di, B., Song, L., Li, Y., and Li, G.Y. (2017). Non-orthogonal multiple access for high-reliable and low-latency V2X communications in 5G systems. *IEEE Journal on Selected Areas in Communications* 35 (10): 2383–2397.

30 Tabassum, H., Hossain, E., and Hossain, J. (2017). Modeling and analysis of uplink non-orthogonal multiple access in large-scale cellular networks using Poisson cluster processes. *IEEE Transactions on Communications* 65 (8): 3555–3570.

31 Khazali, A., Sobhi-Givi, S., Kalbkhani, H., and Shayesteh, M.G. (2018). Resource allocation and power control for underlay device-to-device communication in fractional frequency reuse cellular networks. *Wireless Networks*. 65: 677–697.

32 Sobhi-Givi, S., Khazali, A., Kalbkhani, H. et al. (2017). Resource allocation and power control for underlay device-to-device communication in fractional frequency reuse cellular networks. *Telecommunication Systems* 65 (4): 677–697.

33 Sobhi-Givi, S., Khazali, A., Kalbkhani, H. et al. (2018). Joint mode selection and resource allocation in D2D communication based underlaying cellular networks. *Telecommunication Systems* 67 (1): 47–62.

34 Ali, M.S., Hossain, E., and Kim, D.I. (2018). Coordinated multipoint transmission in downlink multi-cell NOMA systems: models and spectral efficiency performance. *IEEE Wireless Communications* 25 (2): 24–31.

35 Ali, K.S., Elsawy, H., Chaaban, A., and Alouini, M.-S. (2017). Non-orthogonal multiple access for large-scale 5G networks: interference aware design. *IEEE Access* 5: 21204–21216.

36 Ding, Z., Liu, Y., Choi, J. et al. (2017). Application of non-orthogonal multiple access in LTE and 5G networks. *IEEE Communications Magazine* 55 (2): 185–191.

37 Chen, Z., Ding, Z., and Dai, X. (2016). Beamforming for combating inter-cluster and intra-cluster interference in hybrid NOMA systems. *IEEE Access* 4: 4452–4463.

38 Nguyen, T.H., Nguyen, T.K., Han, H.D., and Nguyen, V.D. (2018). Optimal power control and load balancing for uplink cell-free multi-user massive MIMO. *IEEE Access* 6: 14462–14473.

39 Zhang, L., Liu, J., Xiao, M. et al. (2017). Performance analysis and optimization in downlink NOMA systems with cooperative full-duplex relaying. *IEEE Journal on Selected Areas in Communications* 35 (10): 2398–2412.

40 Ramezani-Kebrya, A., Dong, M., Liang, B. et al. (2017). Robust power optimization for device-to-device communication in a multi-cell network under partial CSI. *IEEE International Conference on Communications (ICC)*, pp. 1–6.

41 Ramezani-Kebrya, A., Dong, M., Liang, B. et al. (2017). Joint power optimization for device-to-device communication in cellular networks with interference control. *IEEE Transactions on Wireless Communications* 16 (8): 5131–5146.

42 Yoon, T., Nguyen, T.H., Nguyen, X.T. et al. (2018). Resource allocation for NOMA-based D2D systems coexisting with cellular networks. *IEEE Access* 6: 66293–66304.

43 Boyd, S. and Vandenberghe, L. (2004). *Convex Optimization*. Cambridge University Press.

44 Chien, T.V., Björnson, E., and Larsson, E.G. (2016). Joint power allocation and user association optimization for Massive MIMO systems. *IEEE Transactions on Wireless Communications* 15 (9): 6384–6399.

45 CVX Research Inc. (2015). CVX: Matlab software for disciplined convex programming, academic users.

46 Horn, R.A. and Johnson, C.R. (2003). *Matrix Analysis*. Cambridge University Press.

6

Distributed Multi-Agent RL-Based Autonomous Spectrum Allocation in D2D-Enabled Multi-Tier HetNets

Kamran Zia[1], Nauman Javed[1], Muhammad N. Sial[2], Sohail Ahmed[3], Asad A. Pirzada[1], and Farrukh Pervez[1]

[1] *School of Electrical Engineering & Computer Science, National University of Sciences and Technology, Islamabad, Pakistan*
[2] *Department of Informatics, King's College London, London, United Kingdom*
[3] *Department of Avionics Engineering, Air University, Islamabad, Pakistan*

The increase in multimedia applications has given rise to the requirement of higher data rates in cellular networks. Applications like internet of things (IoT), machine-to-machine communication, personalized TVs, video streaming, video conference calls, and self-driven cars require network with high bandwidth, data rate, and different latencies for their operation. The fifth-generation networks are envisioned to meet these requirements and provide higher data rates up to tens of Gbps to 1000-fold more devices [1]. Moreover, mobile phone usage has increased many fold over the past few years and is continuously rising. According to a Ericsson Mobility Report 2017, mobile subscriptions have reached 7.79 billion in November 2017 and will rise to 9.12 billion by year 2023 [2]. In order to provide cellular services to such larger number of users, technologies like device-to-device communication and multi-tier heterogeneous networks (HetNets) are being developed and tested. HetNets comprises of micro, pico, and femto base stations operating within the coverage of macro base stations (Figure 6.1). The radio access technology of each base station with asymmetrical transmit powers determine the cell sizes and add to the complex interference scenario in the uplink (UL) and downlink (DL) of the network [3].

Similarly, device-to-device (D2D) communication is a promising technology that can provide higher data rates to the users in dense HetNets. D2D communication technology has become the focus of fifth-generation communication systems to support device-centric proximity-based services, i.e. offloading calls to users located in close proximity, context-aware applications, group multi-casting, multimedia data sharing, multi-player online gaming, and public safety through emergency messaging, etc. The close proximity communication provides significant benefits in terms of throughput and hop gain, thus lowering latency. D2D communication can take place in either dedicated mode where base station assigns orthogonal frequency resources with D2D peers for communication or it can take place in shared mode in

Interference Mitigation in Device-to-Device Communications, First Edition.
Edited by Masood Ur Rehman, Ghazanfar Ali Safdar, and Mohammad Asad Rehman Chaudhry.

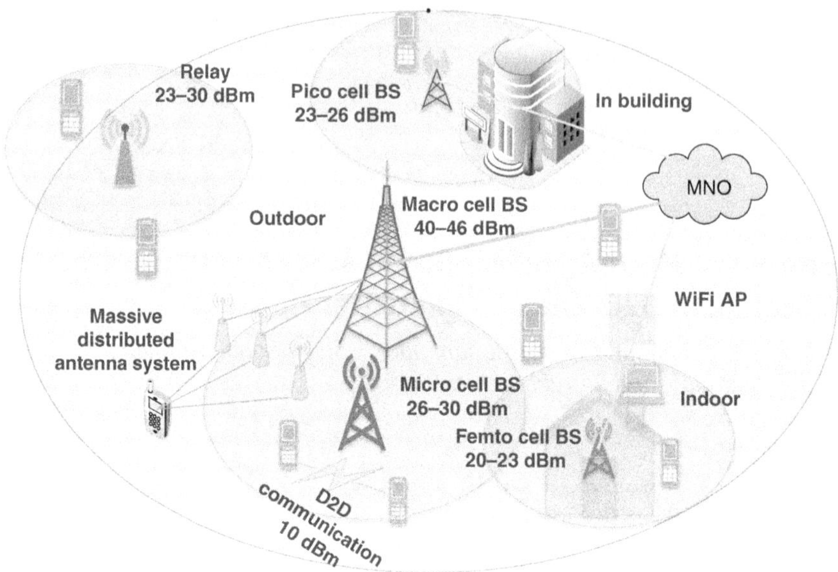

Figure 6.1 Infrastructure of heterogeneous network.

which it reuses frequency spectrum being used by conventional cellular users. Due to the increase in cellular user density, the networks are facing spectrum scarcity; therefore, performance benefits of D2D communication can be best achieved if it takes place in shared mode, also known as underlay mode [31]. The D2D-enabled HetNets can, however, induce significant challenges in terms of co-tier and cross-tier interference, and resource allocation becomes quite challenging [4].

6.1 D2D Resource Allocation Methods

Different techniques have been employed in literature for resource allocation to D2D users operating in underlay mode. Author in [5] proposed distance-based resource criterion (DRC) for sharing cellular resources with D2D users in single-cell network. Although, this technique gives significant results in terms of throughput and outage of cellular users; however, it causes significant overhead to the base stations. Similar techniques are presented in [6] and [7], but they either require network assistance or cause significant overhead to the base stations. Ant colony optimization (ACO)-based resource allocation is proposed in [8]. Graph coloring is used for mapping the interference among D2D users using interference-level indicator (ILI) term. Afterward, ACO algorithm is employed to determine the optimum resource sharing among D2D and cellular users. Authors in [9] have proposed a network-assisted interference mitigation scheme in which D2D users autonomously select resource blocks (RBs). The minimum transmission power level for D2D users is calculated based on the information received from base stations for all RBs and those RBs, which cause tolerable interference to macro and femto users, are selected. In order to reduce the overhead to the macro base stations, different machine learning-based techniques are also proposed in the literature.

Machine learning has been considered as a powerful tool in solving different network problems in fifth-generation networks [10, 11]. Regression model, K-nearest neighbors (KNN), and support vector machines (SVM) algorithms are employed in solving massive multiple-input multiple-output (MIMO) channel estimation/detection and user location/behavior classification problems. Similarly, Bayesian learning is employed in spectrum sensing and detection, K-means clustering is employed in small cell and HetNets clustering and device-to-device user clustering problems, principal component analysis (PCA) has been employed in anomaly, fault and intrusion detection problems, and markov decision process (MDP), multi-armed bandits (MAB), and reinforcement learning (RL) have been employed in decision-making under unknown channel conditions and spectrum sharing for D2D networks. The employment of machine learning in 5G is shown in Figure 6.2.

RL has been greatly employed in providing solution to the resource allocation problems in 5G networks. A Q-learning-based resource allocation is proposed in [12] in a single-tier network. Resources are shared among D2D and cellular users using Q-learning-based strategy to maximize the network throughput. Authors in [13] proposed an expected Q-learning-based resource allocation for LTE-U networks with downlink–uplink decoupling (DUDe) technique. The spectrum allocation problem is formulated as game-theoretic model which incorporates load balancing, spectrum allocation, and user association. The resources are allocated by base stations in autonomous manner with only limited information about the network. Authors in [14] have proposed a Q-learning-based resource allocation scheme in two-tier network. The D2D users and femto cell user determine the RBs

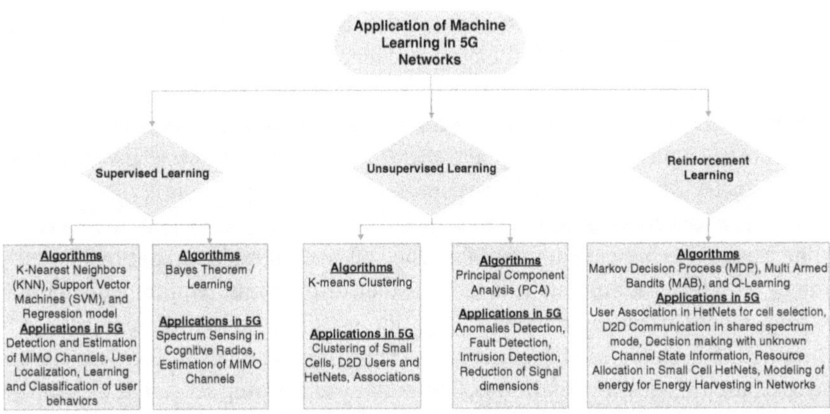

Figure 6.2 Machine learning in 5G.

in cooperative manner for meeting their quality of service (QoS) requirements. The devices learn the optimal strategy by sharing learnt information to select RBs by taking actions (selecting RB) and measuring the feedback of actions in the form of rewards (QoS requirements). In this way, the users select RB in autonomous manner and adapt to the changing network conditions as well. However, the network model considered has single macro base station and only one user associated with each femto base station, which is not practical. Moreover, sharing of information uses additional network resources, which is quite expensive. Authors in [15] proposed a genetic algorithm-based joint resource allocation and user matching scheme to maximize the network throughput. The frequency resources are firstly allocated to the cellular users and then shared by the D2D users in a coordinated manner. Afterward, optimal user matching is performed using genetic algorithm to maximize the system throughput.

The task of interference mitigation can be solved through power control techniques, fractional frequency reuse approach, and resource allocation strategies. The fractional frequency reuse (FFR) scheme [16] divides the region based on frequency selection as inner region and outer region and allocate resources to D2D users based on their location in certain region. Different power control schemes for single-tier network have been proposed to mitigate interference among D2D and cellular users including target signal-to-interference ratio (SIR) tracking power control [17], and opportunistic power control (OPC) [18]. Although, these power control techniques are distributive in nature; however, they do not address the problem of interference in multi-tier HetNet.

In this chapter, we propose a distributive RL-based resource allocation scheme in multi-tier HetNet to mitigate interference between D2D users and cellular users. This work is based on [19] where a learning-based autonomous spectrum allocation (ASA) scheme for two-tier HetNets was found to enhance throughput and spectral efficiency for the D2D users while meeting QoS requirements of conventional cellular users. In this work, we have considered multi-tier HetNets and analyzed the effects of increase in base station density and network tier over the performance of D2D users. Moreover, fairness of spectrum allocation among D2D users, additional memory requirements of the proposed algorithm, and its computational time efficiency compared to other techniques are also taken into account. The discussed scheme focuses on user-centric decision-making for efficient spectrum selection by D2D users in a realistic and more practical stochastic geometry-based multi-tier HetNet. Enhanced inter-cell interference coordination (eICIC) scheme is used to control the interference between multiple tiers of the network; however, D2D users employ online learning to mitigate interference. The D2D users employ an optimal learning strategy to determine suitable RBs to meet their throughput requirements without the prior model of the environment. The prior model of the environment cannot be achieved due to unplanned deployment of femto base stations in the network and due to changing network conditions. The employed learning scheme is distributive in nature; therefore, it reduces the complexity of system implementation. Moreover, it waves off the excessive overhead to the base stations, which is present in case of centralized scheduling of resources.

6.2 Reinforcement *Q*-Learning

The employed learning algorithm utilizes reinforcement *Q*-learning algorithm to determine the optimal policy π^* for taking decisions without the availability of perfect model of the environment. The cellular network environment keeps on changing with time and therefore it cannot be modeled. In order to determine the optimal policy for such a system, *Q*-learning is a useful tool for decision-making. *Q*-learning measures the quality of its action through feedback in the form of rewards and improves them without taking into account the factors affecting performance of the network-like conditions of channel and user mobilities, etc. *Q*-learning comprises of four parameters namely action a, state s, state transition probability $P_{s,s'}$, and reward $R_{s,a}$. The *Q*-learning algorithm runs on an agent that learns the environment through interaction with the environment and taking feedback of its actions. The state of an agent is its internal present condition or it may be external to agent such as usage of a particular spectrum resource. The reward is the measure of the feedback of the action that an agent takes in particular state. Consequently, the agent learns the environment and gains experience. *Q*-learning algorithm works as follows: the agent takes an action a_t in state s_t at some time instant t. As a result of the action a_t, the agent transitions to a new state according to $P_{s,s'}$ and feedback the measured reward $R(s_t, a_t)$ to the agent. The process is repeated for the next state in order to find the optimal *Q*-value for each pair of states and actions. The purpose of finding optimal *Q*-value is to determine the optimal strategy π_s^*. The optimal *Q*-value is determined as follows:

$$Q^*(s,a) = E\left[R(s,a) + \gamma \sum_{s \in S} P_{s,s'}(a) \max_{b \in A} Q^*(s,b)\right] \tag{6.1}$$

where A and S represents the set of states and actions of the learning agent and γ represents the discount factor. The optimal policy is determined as follows:

$$\pi_s^* = \arg \max_{a \in A} Q^*(s,a) \tag{6.2}$$

The employed learning algorithm, however, determines the optimal *Q*-value in a recurring manner as follows:

$$Q(s,a) = (1-\alpha)Q(s,a) + \alpha\left[R(s,a) + \gamma \max_{b \in A} Q^*(s',b)\right] \tag{6.3}$$

where α is the learning rate of the agent. The appropriate actions are rewarded by increasing their *Q*-values, while inappropriate actions are punished and their *Q*-values are decreased. The *Q*-values are stored in a two-dimensional matrix consisting of state versus action entries. The convergence of this algorithm to an optimal *Q*-value has been proven in [20], provided that the states and actions are visited infinitely often. Therefore, the *Q*-learning algorithm employs two stages, namely exploration and exploitation. In exploration, the agent explores different actions chosen randomly to discover highly rewarded actions while in exploitation, the agent chooses best actions in a given state according to learnt policy π_s*. The level of exploration and exploitation is balanced through different techniques like ϵ-greedy action selection or softmax selection.

6.3 System Model

We have considered a network model consists of a multi-tier HetNet with macro, micro, pico, and femto base stations and D2D users who operate in underlay mode. The base stations are spatially distributed in R^2 according to a 2D Poisson point process (PPP) φ_n with intensity λ_n where $n = \{1, 2, \dots, N\}$. The cellular users are spatially distributed with intensity λ_u in R^2 according to 2D-PPP φ_u. The base stations are employing orthogonal frequency-division multiple access (OFDMA) scheme for distributing resources to the cellular users associated with them. Due to OFDMA scheme, there will be only one interferer in the coverage of each cell. We assume that D2D users have already been discovered in the network. The D2D transmitters are uniformly distributed in the area R^2 and D2D receivers are lying in the isotropic region of radius r_d around the D2D transmitters. The distance between a particular D2D transmitter and its intended receiver is assumed to be uniformly distributed in the isotropic region. This assumption is of practical interest and has been considered in many works [21–23]. Other distributions of the distance between D2D transmitter and receiver can easily be incorporated. A half-duplex D2D communication mode is considered employing single-carrier frequency-division multiple access (SC-FDMA) scheme. Since the devices are communicating directly in D2D communication therefore, energy efficiency is an important constraint to save devices power. Due to low peak-to-average power ratio (PAPR) in SC-FDMA mode [24], D2D users save energy to increase battery timings. The network model is illustrated in Figure 6.3.

The D2D users operate in underlay mode sharing same spectrum resources as used by the cellular users in coverage of macro and small-cell base stations; therefore, they

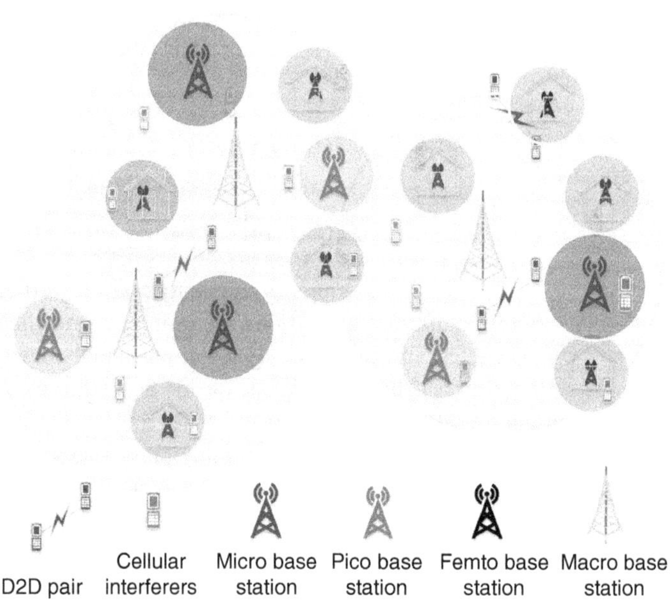

Figure 6.3 Illustration of network model.

will cause interference to cellular users. The interference to the D2D users operating on a particular channel will come from one cellular user from each cell because of OFDMA scheme employed by them. Each D2D user from set of $\{1, 2, \dots, D\}$ users will select one resource from set of $\{1, 2, \dots, N\}$ spectrum resources available to them in an autonomous manner to maximize their throughput (TH_{D_n}) and keep the interference to the cellular user under limits such that signal-to-interference plus noise ratio (SINR) of cellular user does not fall below a minimum required $\zeta_{\min}$. The objective of D2D users is to maximize their throughput while keeping the SINR of cellular user above required threshold.

We have considered a general path-loss power model in which the signal power decays at the rate $r^{-\beta}$ with the propagation distance r. The path-loss exponent is considered same for all the base stations and users in the network. Rayleigh fading has been considered to account for the channel fading, where its channel gain is assumed to be exponentially distributed with unit mean. Different notations used in our network model are listed in Table 6.1.

The cellular users are associated with the base stations based on maximum received signal strength, i.e. the users will associate with the base station from which it receives the maximum power in downlink. We have considered uplink spectrum sharing by D2D users with the cellular users. The small-cell base stations also operate in underlay mode and utilize the same frequencies as used by the macro base stations. This will cause severe interference to the macro-cell users, thus causing inter-cell interference. Several techniques have been introduced in LTE-A under the name enhanced inter-cell interference Coordination (eICIC) [25].

Table 6.1 Notations.

S. No.	Notations	Definitions
1	λ_n	Intensity base stations, 2D PPP
2	λ_u	Intensity of cellular users, 2D PPP
3	φ_n	Poisson point process for base stations
4	φ_u	Poisson point process for cellular users
5	β	Path loss exponent
6	P_m	Transmission power of macro BS
7	P_{mi}	Transmission power of micro BS
8	P_p	Transmission power of pico BS
9	P_f	Transmission power of femto BS
10	P_u	Transmission power of cellular users
11	P_d	Transmission power of D2D transmitter
12	r_d	Distance between D2D pair
13	TH_{D_n}	Throughput of D2D user
14	$\zeta_{\min}$	SINR threshold for cellular users
15	θ	Intercell interference factor

In this paper, we have assumed that eICIC has already been employed by the base stations to mitigate the inter-cell interference; therefore, our purpose is to mitigate interference caused by D2D users to the cellular users in shared mode. For inter-cell interference, we have assumed an inter-cell interference factor θ to denote the virtual reduction of interference among different cells, and this factor θ is constant throughout the network. Therefore, the transmission power of the interferers will become

$$P_I = P_u * \theta \tag{6.4}$$

RL is employed in a distributed manner where each D2D user learns its environment to choose the best spectrum resource for communication to maximize its throughput while keeping interference to the cellular users under control. The learning is done in a recursive manner to determine a policy π^* to find the best available spectrum resource for communication.

6.4 Resource Allocation in Multi-tier D2D Communication

The resource allocation problem in our multi-tier network model with D2D users underlaying the uplink cellular spectrum has the objective of maximizing the D2D users throughput with minimal interference to the cellular users to keep their SINR above certain threshold ($\zeta_{\min}$). SINR of D2D receivers and SINR of cellular users are the main parameters that control the achievement of this objective. In order to calculate the SINR of D2D users, we need to find the sources of interference to the D2D receiver D_r communicating with its transmitter using some RB n from the set of $\{1, 2, \dots, N\}$ RBs. As the base stations are employing OFDMA scheme, therefore, the D2D user will experience interference from one cellular user residing in each cell operating on the same RB n. The interference experienced by the D2D receiver D_r is given by:

$$I_{D_r} = \sum_{x_n \in X} P_I G(x_n, D_r) + \sum_{D_n \in D} P_d G(D_n, D_r) \tag{6.5}$$

where D_r is the underlay D2D receiver which is experiencing interference, x_n are the cellular users operating on RB n in each cell, P_I is the transmit power of cellular users, and $G(x_n, D_r)$ is the interference gain of cellular user x_n toward D2D receiver D_r. Similarly, P_d is the transmit power of D2D transmitters, D_n are the D2D transmitters operating on RB n, and $G(D_n, D_r)$ is the interference gain of D2D transmitters D_n toward D2D receiver D_r. The D2D transmitters are sharing the uplink resources of cellular users; therefore, the interference experienced by D2D receivers will come from the cellular users, not from the base stations. The interference gain of certain cellular user x_n or D2D transmitter D_n toward D_r over RB n is given by:

$$G(x_n, D_r) = F_{x_n,D_r} d_{x_n,D_r}^{-\beta} \tag{6.6}$$

where F_{x_n,D_r} is the channel fading component between x_n and D_r using RB n, d_{x_n,D_r} represents the distance between x_n and D_r, and β represents the *path-loss exponent*. The SINR of the D2D receiver D_r over RB n is given by:

$$\zeta_{D_u} = \frac{P_d G(D_t, D_r)}{I_{D_r} + \sigma^2} \tag{6.7}$$

where ζ_{D_u} is the SINR of D2D receiver D_u, P_d is the transmit power of D2D transmitter D_t, and $G(D_t, D_r)$ represents gain between D_t and D_r. σ^2 is the noise calculated by $\sigma^2 = N_o \mathrm{BW}_{\mathrm{RB}}$ where N_o is the thermal noise and $\mathrm{BW}_{\mathrm{RB}}$ is the bandwidth of the RB. In similar manner, the SINR of the conventional cellular user can also be calculated and is given by:

$$\zeta_{c_n} = \frac{P_u G(x_n, c_n)}{\sum\limits_{D_n \in D} P_d G(D_n, c_n) + \sum\limits_{x_n \in X} P_I G(x_n, c_n) + \sigma^2} \tag{6.8}$$

where ζ_{c_n} is the SINR of cellular user c_n operating on RB n, and $G(D_n, c_n)$ and $G(x_n, c_n)$ are gains of D2D transmitters and cellular users in the direction of c_n, respectively. $P_I = P_u\theta$ is the transmission power of the interferers incorporating the eICIC factor θ as described in section III. Note that the interference to the cellular user c_n is coming from other cellular users and D2D transmitters, operating on same RB n because of uplink resource sharing by D2D users. If the D2D users share downlink resources with the cellular users, then the interference to cellular user will come from base stations and D2D transmitters over particular RB. Our analysis and scheme is, however, focused on UL resource sharing.

In order to calculate the throughput of D2D users and cellular users, SINR is required which can be calculated using (6.7) and (6.8), respectively. The throughput of a particular D2D user D_n over RB n can be calculated by Shannon formula as follows:

$$\mathrm{TH}_{D_n} = \mathrm{BW}_{\mathrm{RB}} log_2(1 + \zeta_{D_n}) \tag{6.9}$$

The selection of a RB n by a D2D user D_n in order to maximize its throughput is represented by a binary decision variable Z_{D_n} where

$$Z_{D_n} = \begin{cases} 1 & \text{if } D_n \text{ is transmitting over RB } n \\ 0 & \text{otherwise} \end{cases} \tag{6.10}$$

The achievable throughput of the D2D user D_n can, therefore, be written as:

$$\mathrm{TH}_{D_n} = \sum_{n=1}^{N} Z_{D_n} \mathrm{BW}_{\mathrm{RB}} \log_2(1 + \zeta_{D_n}) \tag{6.11}$$

In order to maintain the SINR of the cellular user above certain threshold ($\zeta_{\min}$), the interference caused by D2D users sharing uplink spectrum resources should be kept below the threshold of maximum tolerable interference by the cellular users I_{TH} as follows:

$$I_{D_n} = \sum_{D_n=1}^{D} Z_{D_n} P_d G(c_n, D_n) \leq I_{\mathrm{TH}} \tag{6.12}$$

The objective of our spectrum allocation scheme is to maximize the throughput of the D2D users while maintaining the SINR of cellular users above $\zeta_{\min}$. Therefore, our optimization problem can be expressed as follows:

$$\max_{n \in N} \sum_{n=1}^{N} Z_{D_n} \mathrm{BW}_{\mathrm{RB}} \log_2(1 + \zeta_{D_n}) \tag{6.13}$$

subjected to

$$\begin{aligned} \text{C.1 } I_{D_n} &\leq I_{\mathrm{TH}} \quad \forall n \in N, D_n \in D \\ \text{C.2 } Z_{D_n} &\in \{0,1\} \quad \forall n \in N, D_n \in D \\ \text{C.3 } \zeta_{c_n} &\geq \zeta_{\min} \quad \forall n \in N, c_n \in C \end{aligned}$$

where ζ_{c_n} is presented in (6.8). The problem of allocation in (6.13) tries to maximize the throughput of the D2D users while fulfilling the constraints defined in C.1, C.2, and C.3. The interference caused by the D2D users to the cellular users while reusing the cellular uplink resources is limited by a predefined tolerable interference threshold defined by constraint C.1. C.2 puts the restriction that each D2D user can use only one RB at a time and C.3 defines that the SINR of the cellular user remains above the predefined SINR threshold ($\zeta_{\min}$). Note that the C.3 is supported by the assumption that base stations can exchange their associated cellular users SINR with the D2D transmitters as they operate under their coverage [26].

6.4.1 Autonomous Spectrum Allocation Scheme

We have considered a multi-agent RL-based system in which each D2D user acts as a learning agent. The agent collects the environment information from its own and define its state s_t at time t and performs an action a_t. Due to this action, the agent moves to new state and measures the reward $R(s_t, a_t)$ for the action. The agent calculates Q-value for the state s_t and action a_t using (6.3) and records it in its Q-table. The process is repeated with new state and after certain time, the agent populates its Q-table with enough information to devise a strategy π^* to select RBs from set of $\{1,2,\ldots,N\}$ RBs in order to maximize its throughput while meeting all the constraints. Each agent (D2D user) has a role of learning agent, which tries to reach an optimum strategy for RB selection in different network states. The basic parameters of our learning system are as follows:

- *State*: The learning agents rely on their locally observed environment conditions to define their state at a particular time slot t because of no cooperation among themselves. The state observed by agent i is defined as:

$$s_i^t = (i, n_i) \tag{6.14}$$

  where n_i represents the RB selected by agent i.
- *Action*: The agent has N number of RBs available to it for communication; therefore, the action of an agent is the selection of particular RB n i.e. $(a_i) = (n_i)$.

- *Reward*: The reward function is defined on the basis of throughput achieved by the agent for each state–action pair as $R_i(s_t, a_t)$ and is given by:

$$R_i(s_t, a_t) = \begin{cases} R_t(a_t) = \text{TH}_{D_n} & \text{if C.1 to C.3 satisfied} \\ -100 & \text{otherwise} \end{cases} \tag{6.15}$$

Our learning system exploits the collected state–action Q values to determine the optimal strategy for spectrum allocation. The learning of the agents is divided into two stages, namely exploration and exploitation. The agent continuously explores different actions in different environment states and updates its Q-table in a recursive manner as follows:

$$\begin{aligned} Q_i(s_i, a_i) = {} & (1 - \alpha)Q_i(s_i, a_i) + \alpha[R_i(s_i, a_i) \\ & + \gamma \max_{l \in A_i} Q_i(s_i', l)] \end{aligned} \tag{6.16}$$

where s_i is the current state and s_i' is the next state of the agent i. The learning agent tries to learn a strategy to maximize the future long-term reward. The learning system relies on the SINR and data rates, etc., to take the actions in exploration phase. In order to balance the exploration and exploitation, ϵ-greedy method has been employed. The system starts with high exploration rate ϵ to find maximum actions with high Q-values to exploit. The actions that meet the constraints are rewarded, while others are punished by giving negative reward. In this way, when the network conditions change, the agent automatically adapts to select a new action for meeting the constraints and maximizing its throughput. The selection of actions in the exploitation phase is done as follows:

$$a_i = \arg\ \max (Q_i(s_i, a_i)) \tag{6.17}$$

In our scheme, each agent learns its own strategy to maximize the throughput. The pseudo-code for the employed learning system for ASA by D2D users is given in Table 6.2.

6.5 Performance Evaluation

We have evaluated our ASA scheme for D2D users underlaying UL cellular resources through extensive simulations in MATLAB. The simulations were run in a multi-tier stochastic geometry-based HetNet with multiple macro, micro, pico, and femto base stations, deployed randomly using PPP and are employing OFDMA scheme. Multiple D2D users are uniformly distributed in the network and reuse the UL cellular resources. A set of 15 RBs are distributed to the base stations as well as D2D users, whereas each RB has 12 subcarriers having bandwidth of 180 kHz each. Therefore, each RB has a bandwidth of $12 \times 180 = 2160$ kHz available. The D2D users autonomously select RBs from these 15 available RBs to maximize their throughput while maintaining QoS requirements of cellular users. We simulated the proposed scheme using Monte Carlo technique comprising of 10 000 simulation runs. In order

Table 6.2 Basic working of autonomous spectrum allocation scheme.

Autonomous Spectrum Allocation Scheme
1. Initialize Q-Matrix Q for storing Q-values of the state action pairs
2. Observe the current state s_t of the environment at time t
3. Check ϵ for Exploration or Exploitation
4. if *Exploration* **do**
5. Select action randomly and check the following
6. if
7. Throughput of the D2D user according to (6.11)
8. SINR of cellular users to guarantee its QoS according to (6.8)
9. C.1 and C.2 for fulfillment
10. end if *Exploitation* **do**
11. Select action with the highest Q-value according to (6.17)
12. end
13. Evaluate Reward $R(s_t, a_t)$ for selected action a_t according to (6.15)
14. Update the Q-value in Q-Matrix against state-action according to (6.16)
15. Observe the new environment state s_{t+1} at time $(t + 1)$
16. Repeat this process for next state s_{t+1}

to capture the realistic and dynamically changing network state, the channel fading effects are generated on random basis in each simulation run. The cellular users association with base stations is determined in each simulation run to determine their associated cell IDs which are used to determine their SINR in the uplink direction. The simulation and ASA scheme parameters are chosen as per references and are listed in Table 6.3.

In order to analyze the impact of learning capabilities of the D2D users on their performance metrics (throughput and spectral efficiency), we have compared the evaluation results of our scheme with DRC scheme [5] and Joint-RALA scheme [27] which are separately implemented in MATLAB under identical circumstances in terms of users and network. The DRC scheme utilizes the distance between D2D users and cellular users to determine a RB allocation to the D2D users. The processing is centrally done by base stations and the GPS locations of all cellular users, and D2D users are required by the base stations. Although, DRC gives significant performance in terms of throughput; however, practical implementation of this technique is not possible due to very large overhead of calculating distances in dense 5G HetNets. Moreover, a lot of cooperation will be required to employ this technique in multi-tier HetNet. Joint-RALA scheme proposes a combined algorithm for resource allocation and link adaptation, and it also supports functionality of carrier aggregation in 5G network.

Table 6.3 Network model and simulation parameters.

Simulation parameters used	Value
Bandwidth of system, BW	30 MHz
Resource blocks, N	15
Macro BS Tx power, P_m	46 dBm
Micro BS Tx power, P_{mi}	26 dBm
Pico BS Tx power, P_p	23 dBm
Femto BS Tx power, P_f	20 dBm
D2D Tx power, P_d	10 dBm
Cellular user Tx power, P_u	20 dBm
Macro BS intensity, λ_m	2×10^{-6}
Micro BS intensity, λ_{mi}	6×10^{-6}
Femto BS intensity, λ_p	8×10^{-6}
Femto BS intensity, λ_f	1×10^{-5}
Cellular user intensity, λ_f	6×10^{-4}
No of D2D pairs, D	Variable
eICIC factor, θ	−20 dB
Path loss exponent, β	4
Thermal noise density	−174 dBm/Hz
ASA scheme parameters	**Value**
Learning rate, α	0.2
Discount factor, γ	0.8
Exploration rate, ϵ	Dynamic

6.5.1 Performance of D2D Users

The performance of our scheme is compared with others in terms of aggregate throughput for D2D users (Figure 6.5), Spectral efficiency in bits/Hz (Figure 6.6), and cumulative distribution function (CDF) of average throughput of D2D users (Figure 6.7). Moreover, the fairness of our scheme is determined using Jane's Fairness Index calculated by $f(x_1, x_2, \dots) = \frac{\left(\sum_{i=1}^{J} x_i\right)^2}{J \sum_{i=1}^{J} x_i^2}$ and is shown in Figure 6.4. Our scheme achieves good level of fairness among multiple D2D users by enabling all users to achieve same average throughput. This is because of the learning capability of each D2D users in the network who are populating their own Q-tables for intelligent selection of RBs. The fairness plot is clearly indicating the fair distribution of network resources among D2D users which in one of the major requirements in resource allocation problems.

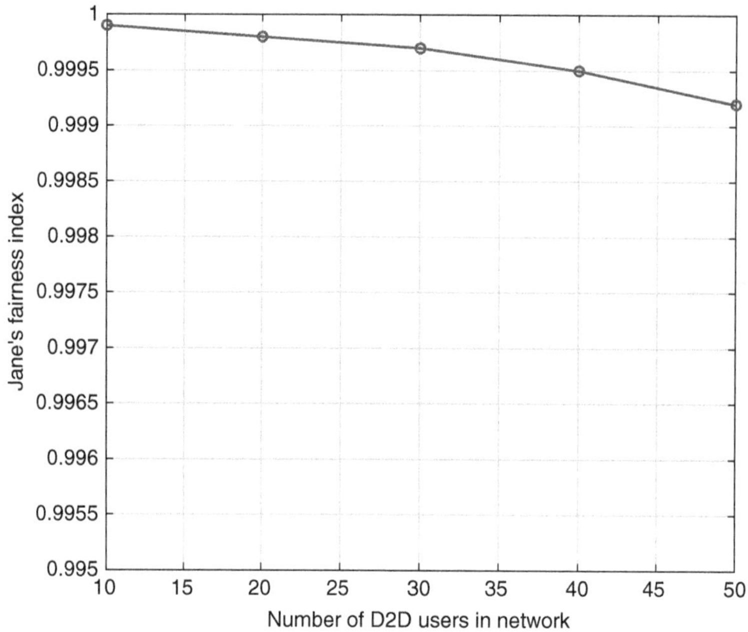

Figure 6.4 Jane's fairness index for ASA scheme.

Figures 6.5 and 6.6 show the plots of aggregate throughput of D2D users against number of D2D users in the network and spectral efficiency for different schemes. A higher aggregate throughput and spectrum efficiency are achieved by our scheme compared to other techniques. This is because our scheme devises the spectrum allocation problem with the objective of throughput maximization and makes use of distributed *Q*-learning as a tool to achieve this objective. As the network conditions change at every moment, the decision making needs to adapt to these changing network conditions. Our system undergoes plenty of interactions with the environment; therefore, it is able to adapt to changing conditions and achieves higher throughput and spectral efficiency.

The CDF plots of average throughput for different schemes in Figure 6.7 also indicate the achievement of higher average throughput by our scheme. The steepness in CDF plot for our scheme shows that D2D users are getting throughput higher than 10–15 Mbps all the time, while throughput of less than 10 Mbps are achieved by other schemes in certain cases.

6.5.2 Performance of Cellular Users

The effectiveness of resource allocation scheme cannot be validated if the performance metrics of conventional cellular users are not met. The key performance indicators (KPI) of cellular users in the network include their SINR and outage ratio. Therefore, the mean SINR of conventional cellular users in the network and outage ratios for different number of D2D users in network is plotted in Figures 6.8 and 6.9,

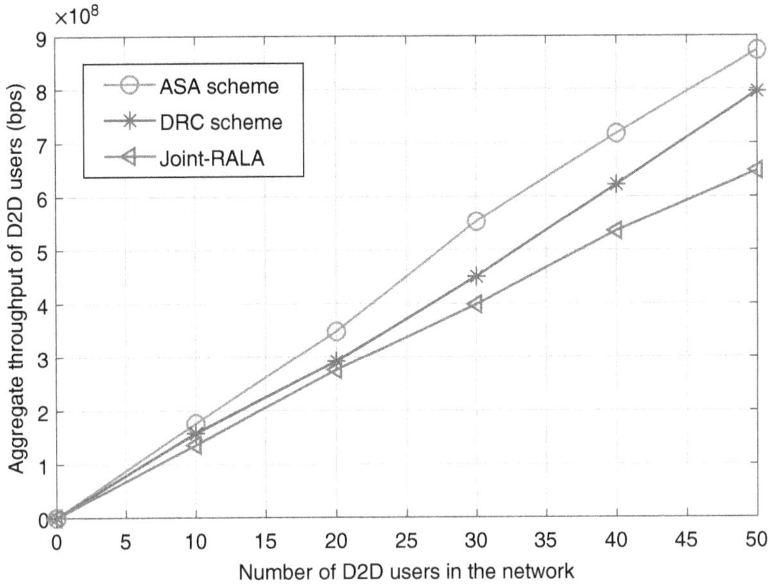

Figure 6.5 Aggregate throughput of D2D users against number of D2D users in the network.

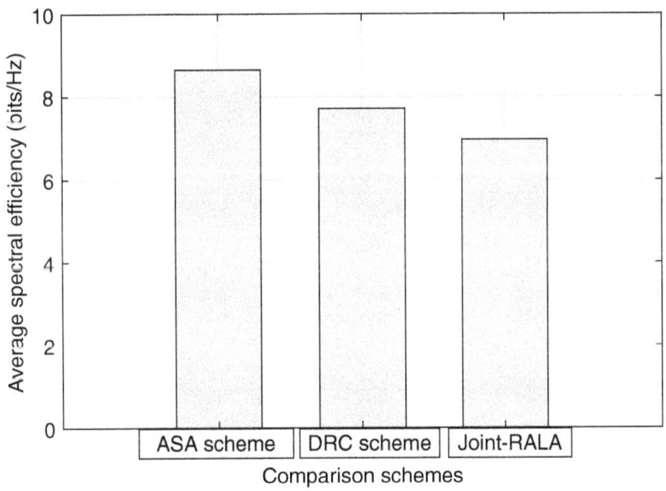

Figure 6.6 Spectral efficiency comparison of different schemes.

respectively. It is clear from Figure 6.8 that the mean SINR of cellular user in our scheme is higher compared to other schemes. This is because our scheme allocates spectrum while causing minimum interference to cellular users and keeps the SINR of cellular user above threshold ζ_{min}. The control over the interference to the cellular user is maintained through the assignment of the rewards to the learning agents (D2D Users). The RBs selected by D2D users which affect the SINR of cellular users

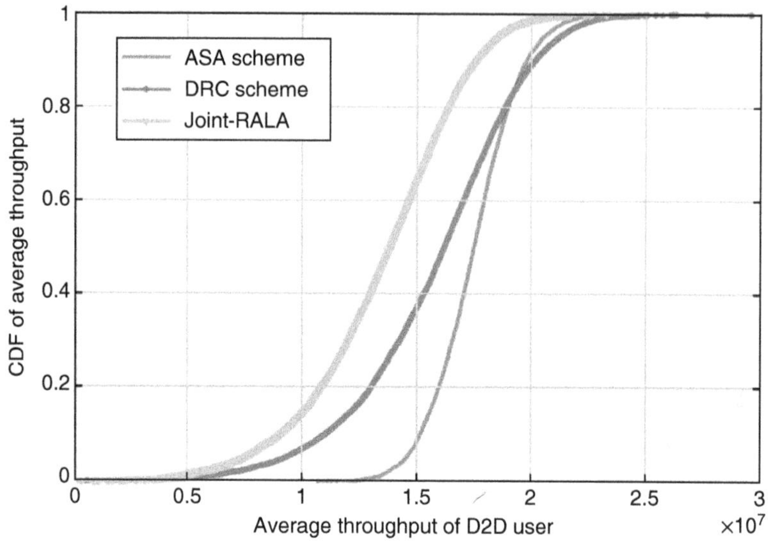

Figure 6.7 Comparison of CDF of average throughput of D2D user in network.

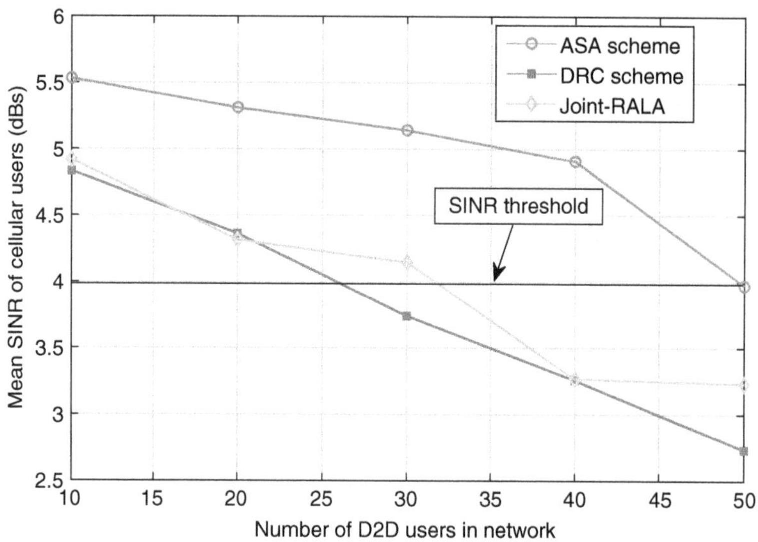

Figure 6.8 Mean SINR of conventional cellular users in network against number of D2D users in the network.

below ζ_{min} are assigned negative rewards to avoid the selection of such RBs. DRC and Joint-RALA schemes do not ensure the ζ_{min} of cellular users and, therefore, achieves lower mean SINR in the network. Due to the control over the interference caused to cellular user in our scheme, it achieves better outage ratio compared to other schemes as shown in Figure 6.9.

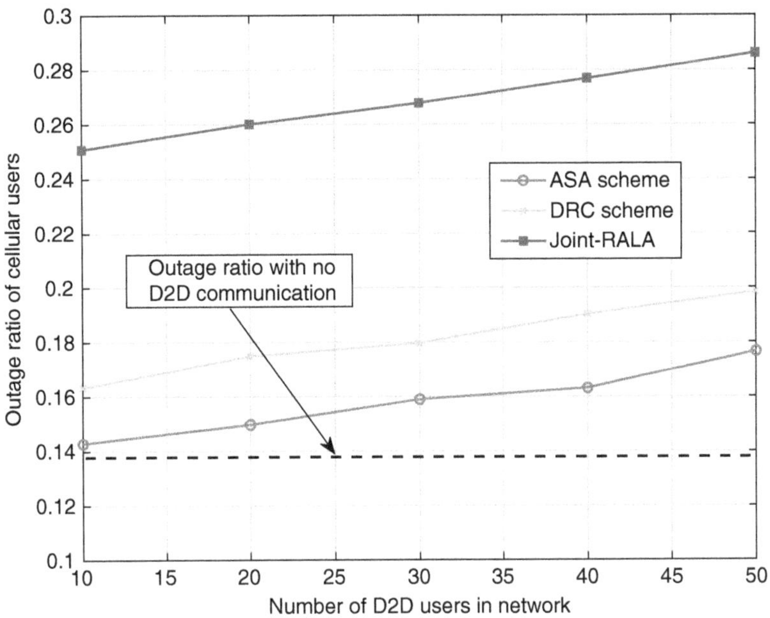

Figure 6.9 Outage ratio of conventional cellular users against number of D2D users in the network.

6.5.3 Coverage Analysis

According to Andrews et al. [28], the coverage probability is simply Pr [SINR > ζ_{min}], where ζ_{min} is the minimum SINR threshold required by the cellular user. The SINR-based coverage probability calculation is a good way to measure the performance of the network at the physical layer. The SINR-based coverage analysis has been considered by many authors in their analysis [29, 30]. We have also performed SINR-based coverage analysis of our network and in order to visualize the coverage, the cumulative distribution function of the coverage of our network is plotted for different schemes. The plots are taken with no D2D user in the network and with 50 D2D users in the network sharing the uplink resources. The objective of this analysis is to find the effect of spectrum sharing by D2D users over the coverage of the network. It can be seen in Figures 6.10–6.12 that our scheme causes minimum reduction in the coverage of the network by underlaying 50 D2D users in the network. This is because of the interference control mechanism employed in our scheme to meet QoS requirements of the cellular users.

6.5.4 Computational Time Analysis

Spectrum scarcity has tunneled the research in the field of D2D-enabled HetNets to spectrum sharing-based solutions. In order to allocate resources to the D2D users sharing resources with the cellular users, the decision time is of considerable

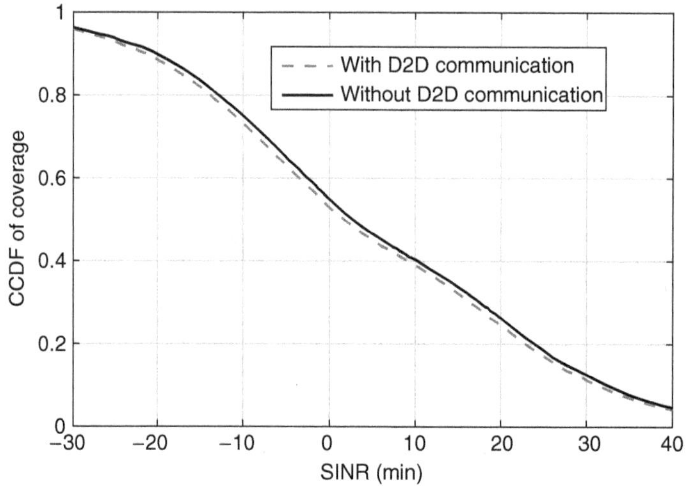

Figure 6.10 CCDF of coverage of network with 50 D2D users-ASA scheme.

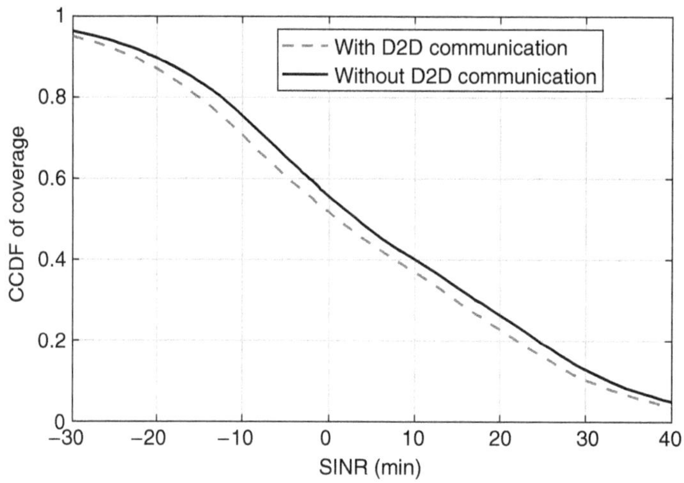

Figure 6.11 CCDF of coverage of network with 50 D2D users-DRC scheme.

importance. A longer decision time for spectrum selection will ultimately lead to lower throughput by the D2D users due to increased latency. Our scheme takes the decision of RB selection in considerable less time compared to other techniques and outperforms them in terms of throughput and spectral efficiency; please refer to Figure 6.13. This is because, the resource selection decisions have been distributed and taken by the D2D users themselves unlike the decisions taken by base station centrally in other schemes. The centralized resource allocation schemes [5, 27] are computationally more extensive and therefore takes more time in determining optimal resource for D2D users. This not only uses more computational power but also causes significant overhead to the base stations. Contrary to this, our

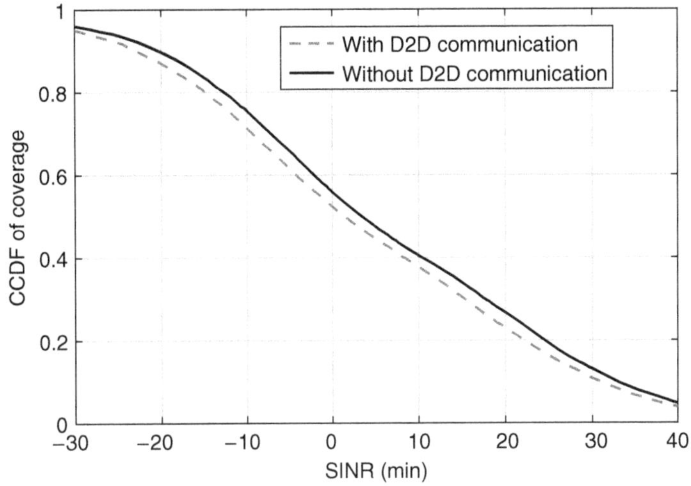

Figure 6.12 CCDF of coverage of network with 50 D2D users-joint RALA scheme.

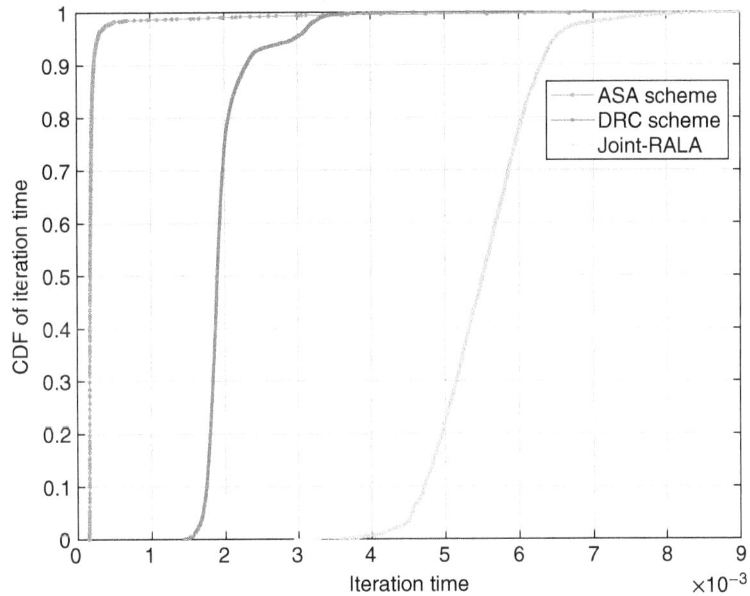

Figure 6.13 Computation time comparison of different schemes for resource allocation (50 D2D users in network).

distributed scheme is computationally more efficient both in terms of computation power and computation time.

6.5.5 Memory Requirements

Look-up tables in the form of Q-tables are an essential component of the reinforcement Q-learning algorithm in order to determine the optimal strategy through mapping of best state–action pairs for every state. These Q-tables help D2D users in effective decision-making for resource selection. Since memory is a critical resource, it becomes necessary to determine the memory requirements of the learning D2D users. Considering all the values in Q-tables of type double that occupies 8 bytes in memory, the memory space required for our scheme is 1800 bytes = 1.757 KB (15 States, 15 Actions). As each D2D user is separately storing its Q-table in its memory, therefore, each user will be utilizing 1.757 KB of memory space. On the contrary in case of centralized resource allocation, the base stations will be computing the distance and measuring link gains for all the admitted D2D users in the network; therefore, memory requirements for storing and processing this information will be considerably large compared to our scheme. The increase in number of D2D or cellular users will not increase memory requirements of our algorithm, whereas the memory requirements of centralized resource allocation schemes increase with the number of admitted users in the network.

6.5.6 Effect of Base Stations Density

The 5G networks are envisioned to provide network coverage to a larger number of users. This will require ultra-dense networks (UDN) with a larger number of base stations to extend network coverage. Therefore, it is necessary to analyze the effect of increase in base station density on the performance of D2D users in terms of aggregate throughput for the proposed scheme. Figure 6.14 shows the effect of increase in macro base station density on the aggregate throughput of 50 D2D users in our network. The increase in macro base stations density will increase the small-cell base

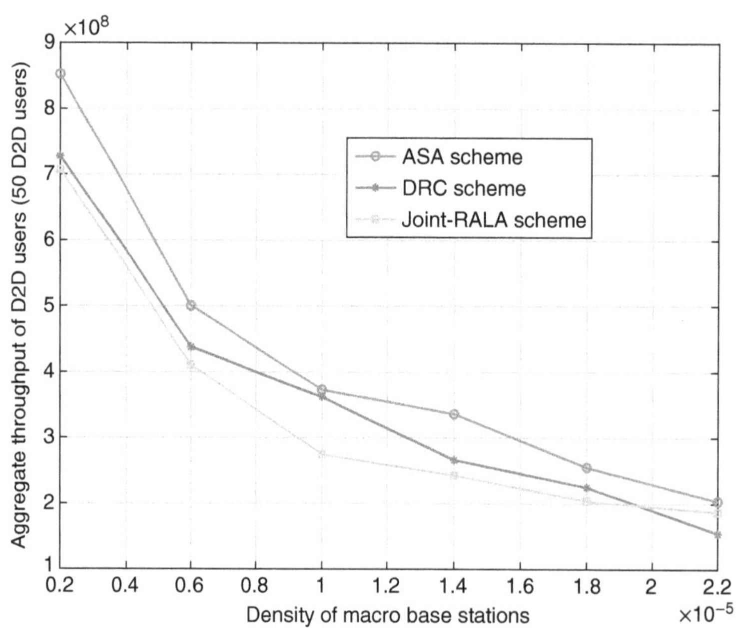

Figure 6.14 Effect of increase in base stations density on aggregate throughput of D2D users in network.

stations density as well due to linear relationship ($\lambda_{\text{mi}} = 3 * \lambda_m$, $\lambda_p = 4 * \lambda_m$, and $\lambda_f = 5 * \lambda_m$) between them. The figure indicates a decrease in aggregate throughput with the increase in base stations density. This is because the increase in base stations causes increase in number of cellular users operating on a particular RB due to OFDMA assumption. This increase in number of cellular users in the network increases the interference over available RBs and therefore reduces the aggregate throughput of D2D users. The D2D users are exploiting spatial reuse of spectrum resources in the network through RL; therefore, increase in cellular users will limit this spatial reuse and hence the aggregate throughput reduces. However, our scheme is still performing better in UDN and enabling D2D users to achieve higher throughput compared to other schemes.

6.5.7 Effect of Network Tiers

Similar to the effect of increase in base station density on the D2D users, we have examined the performance of ASA scheme with increasing network tiers. Two-, three-, and four-tier networks are included in the network model. In three-tier network, macro, micro, and pico base stations are considered. The relationship between macro and micro base stations is kept $\lambda_{\text{mi}} = 3 * \lambda_m$, while relationship between macro and pico base stations is kept $\lambda_p = 4 * \lambda_m$. The transmit power of micro base station is $P_{\text{mi}} = 26$ dBm, while that of pico base station is $P_p = 23$ dBm. Similarly, in four-tier network, we have considered femto base stations as well in addition to macro, micro, and pico base stations. The transmit power of femto base stations is kept 20 dBm in four-tier network. Figure 6.15 shows the effect of tier on performance of D2D users.

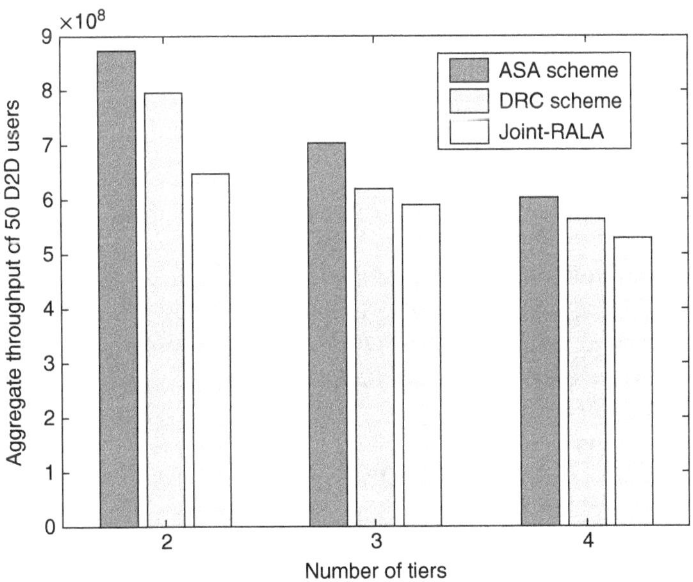

Figure 6.15 Effect of network tier on aggregate throughput of D2D users in the network.

The increase in network tier reduces the aggregate throughput of D2D users. Due to the OFDMA assumption in our network model, there is at least one user in each cell operating on some particular RB; therefore, total number of users operating on a particular RB equals the number of cells or base stations. As the network tier increases, the number of base stations and the cellular interferers over a particular RB will increase causing more interference to D2D users using same RBs. This increased interference limits the spatial reuse of RBs by D2D users thus reducing their aggregate throughput. The comparison plot of effect of tier on D2D users' performance in Figure 6.15 clearly indicates that our scheme outperforms other schemes in multi-tier HetNets as well.

6.6 Summary

In this paper, we have proposed an ASA scheme with distributed Q-learning-based algorithm to allocate spectrum to the underlaying D2D users and control interference in D2D-enabled multi-tier HetNets. Our scheme is distributed and will run on devices to enable them to select spectrum autonomously, thus shedding off the processing load of the base stations. The objective of our proposed scheme is to maximize the throughput of D2D users while maintaining the QoS requirements of the cellular users and cellular outage ratio. The learning methodology employed makes use of interactions with the environment to adapt to changing network conditions in fast and convenient manner to improve the throughput and spectral efficiency. The performance of proposed scheme is compared with other schemes to show its optimality in terms of throughput, spectral efficiency, mean SINR of cellular users, outage ratios coverage of the network, and computation time. Analysis has shown that the proposed technique outperforms other techniques in both UDN and multi-tier HetNets.

References

1 NGMN Alliance (2015). Next generation mobile networks, white paper. 5G white paper.

2 Heuveldop, N. et al. (2017). Ericsson AB. *Tech. Rep. EAB-17. Ericsson Mobility Report. 5964*. Stockholm, Sweden: Emerging Technologies for Business.

3 Olwal, T.O., Djouani, K., and Kurien, A.M. (2016). A survey of resource management toward 5G radio access networks. *IEEE Communication Surveys and Tutorials* 18 (3): 1656–1686.

4 Chin, W.H., Fan, Z., and Haines, R. (2014). Emerging technologies and research challenges for 5G wireless networks. *IEEE Wireless Communications* 21 (2): 106–112.

5 Wang, H. and Chu, X. (2012). Distance-constrained resource-sharing criteria for device-to-device communications underlaying cellular networks. *Electronics Letters* 48 (9): 528–530.

6 Peng, T., Lu, Q., Wang, H. et al. (2009). Interference avoidance mechanisms in the hybrid cellular and device-to-device systems. 2009 IEEE 20th International Symposium on Personal, Indoor and Mobile Radio Communications, pp. 617–621. IEEE.

7 Zulhasnine, M., Huang, C., and Srinivasan, A. (2010). Efficient resource allocation for device-to-device communication underlaying LTE network. 2010 IEEE 6th International Conference on Wireless and Mobile Computing, Networking and Communications (WiMob), pp. 368–375. IEEE.

8 Liotou, E., Tsolkas, D., Passas, N., and Merakos, L. (2014). Ant colony optimization for resource sharing among D2D communications. *2014 IEEE 19th International Workshop on Computer Aided Modeling and Design of Communication Links and Networks (CAMAD)*, pp. 360–364. IEEE.

9 Tsai, A.-H., Wang, L.-C., Huang, J.-H., and Lin, T.-M. (2012). Intelligent resource management for device-to-device (D2D) communications in heterogeneous networks. 2012 15th International Symposium on Wireless Personal Multimedia Communications (WPMC), pp. 75–79. IEEE.

10 Jiang, C., Zhang, H., Ren, Y. et al. (2017). Machine learning paradigms for next-generation wireless networks. *IEEE Wireless Communications* 24 (2): 98–105.

11 Wang, X., Li, X., and Leung, V.C.M. (2015). Artificial intelligence-based techniques for emerging heterogeneous network: state of the arts, opportunities, and challenges. *IEEE Access* 3: 1379–1391.

12 Luo, Y., Shi, Z., Zhou, X. et al. (2014). Dynamic resource allocations based on Q-learning for D2D communication in cellular networks. *2014 11th International Computer Conference on Wavelet Active Media Technology and Information Processing (ICCWAMTIP)*, pp. 385–388. IEEE.

13 Hu, Y., MacKenzie, R., and Hao, M. (2017). Expected Q-learning for self-organizing resource allocation in LTE-U with downlink-uplink decoupling. *23th European Wireless Conference; Proceedings of European Wireless 2017*, pp. 1–6. VDE.

14 AlQerm, I. and Shihada, B. (2016). A cooperative online learning scheme for resource allocation in 5G systems. *2016 IEEE International Conference on Communications (ICC)*, pp. 1–7. IEEE.

15 Yang, C., Xu, X., Han, J. et al. (2014). GA based optimal resource allocation and user matching in device to device underlaying network. 2014 IEEE on Wireless Communications and Networking Conference Workshops (WCNCW), pp. 242–247. IEEE.

16 Chae, H.S., Gu, J., Choi, B.-G., and Chung, M.Y. (2011). Radio resource allocation scheme for device-to-device communication in cellular networks using fractional frequency reuse. *2011 17th Asia-Pacific Conference on Communications (APCC)*, pp. 58–62. IEEE.

17 Foschini, G.J. and Miljanic, Z. (1993). A simple distributed autonomous power control algorithm and its convergence. *IEEE Transactions on Vehicular Technology* 42 (4): 641–646.

18 Leung, K.-K. and Sung, C.W. (2006). An opportunistic power control algorithm for cellular network. *IEEE/ACM Transactions on Networking* 14 (3): 470–478.

19 Zia, K., Javed, N., Sial, M.N. et al. (2018). Multi-Agent RL based User-Centric spectrum allocation scheme in D2D enabled hetnets. *2018 IEEE 23rd International Workshop on Computer Aided Modeling and Design of Communication Links and Networks (CAMAD) (IEEE CAMAD 2018)*, Barcelona, Spain, September 2018.

20 Watkins, C.J.C.H. and Dayan, P. (1992). Q-learning. *Machine Learning* 8 (3–4): 279–292.

21 Ma, C., Wu, W., Cui, Y., and Wang, X. (2015). On the performance of successive interference cancellation in D2D-enabled cellular networks. *2015 IEEE Conference on Computer Communications (INFOCOM)*, pp. 37–45. IEEE.

22 Lin, X. and Andrews, J.G. (2013). Optimal spectrum partition and mode selection in device-to-device overlaid cellular networks. *Global Communications Conference (GLOBECOM), 2013 IEEE*, pp. 1837–1842. IEEE.

23 Lin, X., Ratasuk, R., Ghosh, A., and Andrews, J.G. (2014). Modeling, analysis, and optimization of multicast device-to-device transmissions. *IEEE Transactions on Wireless Communications* 13 (8): 4346–4359.

24 Myung, H.G., Lim, J., and Goodman, D.J. (2006). Peak-to-average power ratio of single carrier FDMA signals with pulse shaping. 2006 IEEE 17th International Symposium on Personal, Indoor and Mobile Radio Communications, pp. 1–5. IEEE.

25 Lopez-Perez, D., Guvenc, I., De la Roche, G. et al. (2011). Enhanced intercell interference coordination challenges in heterogeneous networks. *IEEE Wireless Communications* 18 (3): 22–30.

26 AlQerm, I. and Shihada, B. (2017). Energy-efficient power allocation in multitier 5G networks using enhanced online learning. *IEEE Transactions on Vehicular Technology* 66 (12): 11086–11097.

27 Rostami, S., Arshad, K., and Rapajic, P. (2015). A joint resource allocation and link adaptation algorithm with carrier aggregation for 5G LTE-advanced network. *2015 22nd International Conference on Telecommunications (ICT)*, pp. 102–106. IEEE.

28 Andrews, J.G., Ganti, R.K., Haenggi, M. et al. (2010). A primer on spatial modeling and analysis in wireless networks. *IEEE Communications Magazine* 48 (11): 156–163.

29 Sial, M.N. and Ahmed, J. (2018). Analysis of K-tier 5G heterogeneous cellular network with dual-connectivity and uplink–downlink decoupled access. *Telecommunication Systems* 67 (4): 669–685.

30 Smiljkovikj, K., Popovski, P., and Gavrilovska, L. (2015). Analysis of the decoupled access for downlink and uplink in wireless heterogeneous networks. *IEEE Wireless Communications Letters* 4 (2): 173–176.

31 Safdar, G.A., Ur-Rehman, M., Muhammed, M. et al. (2016). Interference mitigation in D2D communication underlaying LTE-A network. *IEEE Access* 4: 7967–7987.

7

Adaptive Interference Aware Device-to-Device-Enabled Unmanned Aerial Vehicle Communications

Navuday Sharma[1], Rafay I. Ansari[2], Rida Khan[3], Hassan Malik[4], and Haris Pervaiz[5]

[1] *Test Software Development, Ericsson Eesti AS, Tallinn, Estonia*
[2] *Department of Computer and Information Sciences, Northumbria University, Newcastle, United Kingdom*
[3] *Thomas Johann Seebeck Department of Electronics, Tallinn University of Technology, Tallinn, Estonia*
[4] *Department of Computer Science, Edge Hill University, Ormskirk, United Kingdom*
[5] *School of Computing and Communications, Lancaster University, Bailrigg, United Kingdom*

The number of connected devices have grown exponentially, giving rise to new demands in terms of spectrum resources. Fifth-generation (5G) mobile technology envisions efficient utilization of the spectrum and has introduced several enabling technologies such as device-to-device (D2D), Internet-of-things (IoT), and machine-to-machine (M2M) [1–20]. Each technology introduced several advantages in terms of spectrum management and mass connectivity. It is pertinent to note that the development of each technology was based on the key performance metrics defined under 5G, which include the network rates, latency, and energy/spectrum efficiency [21, 22].

D2D technology has emerged as a potential candidate to assist the heterogeneous networks for improving data rates and spectrum efficiency by establishing a direct connectivity between User Equipments (UEs). The major impact of D2D communications is expected in public safety networks, since presence of communication network is not required. In this chapter, we emphasize on utilizing unmanned aerial vehicles (UAVs) to implement D2D network for UEs. On-demand deployment and relays of UAVs can fulfill the data requirements of users as well as establish connectivity for out-of-coverage users. Further, we discuss about Third Generation Partnership Project (3GPP) activities for D2D and show simulation results for throughput and packet error rates (PERs) between UAV and UE D2D link. We also address the consequence of another interfering UAV establishing D2D connection simultaneously with the same UE, by showing variation of PER with interfering UAV power and altitude for different environments.

Interference Mitigation in Device-to-Device Communications, First Edition.
Edited by Masood Ur Rehman, Ghazanfar Ali Safdar, and Mohammad Asad Rehman Chaudhry.

7.1 Key Elements in D2D Communication

D2D communication is based on the idea of exploiting the direct connections between the users instead of relying on the traditional cellular network for communications [23]. The users in close proximity can establish direct links to share the data or realize different multimedia services. The main aim behind D2D communications is to reuse the spectrum resources efficiently as well as to share the burden of data transmissions with the cellular network. The mobility patterns of users can be utilized to create these collaborative D2D links [24]. The cooperative network between devices can help in realizing new services and opening up several new business opportunities. However, there are also several challenges associated with the D2D operation, such as the mobility of users makes the D2D links very dynamic and highlights the need for a comprehensive control mechanism.

The users in proximity can establish links during their interactions but to sustain the transmissions the network requires some intelligence. The required intelligence needs to cater to the mobility patterns and help in the sustainability of the transmissions. Some prediction methods have been proposed to ascertain the frequency of the interaction between devices and develop a cooperative network that can provide a good signaling mechanism. The D2D operation has been analyzed for different mobility models to evaluate its efficacy [25]. D2D communications can greatly help in providing information to users sharing a common interest. For example, the users in a shopping mall can share the information on any new deals or offers. D2D also finds its use in agriculture, health, and sports activities. Depending on the application, different topologies of D2D have been proposed.

Figure 7.1 shows different modes of operation of D2D networks. In-band D2D refers to a mode of operation where D2D users communicate by reusing the licensed cellular spectrum. The in-band D2D is further segregated into underlay and overlay D2D [26]. In underlay D2D, the D2D users reuse the resource blocks assigned to cellular users, while in overlay D2D, the D2D users are assigned dedicated resources. Both underlay and overlay modes have their advantages and disadvantages. For instance, in underlay mode of operation, the spectrum resources are utilized efficiently, but the communication is prone to interference. On the other hand, the overlay mode of operation avoids the interference but leads to inefficient

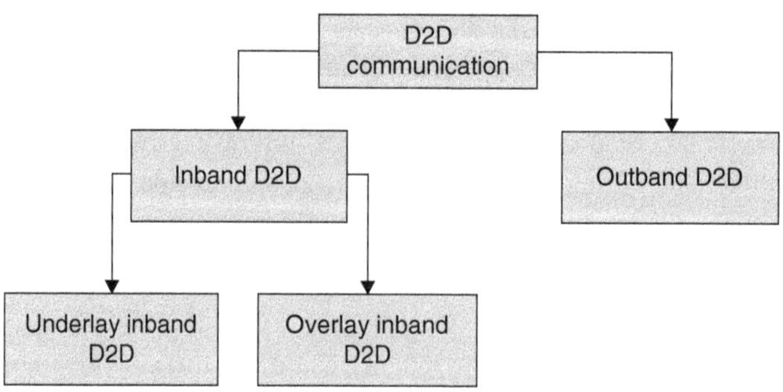

Figure 7.1 D2D modes of operation.

utilization of resources as the dedicated spectrum resources might not be in use by the D2D users, but they cannot be reassigned to cellular users. In the out-band D2D, the D2D users utilize the unlicensed spectrum for communication.

7.1.1 D2D Network Discovery

The D2D communication can only be realized if the devices are able to scan devices in proximity and know the contextual information [10]. However, gathering this information can be taxing for the devices in terms of processing and battery constraints. D2D network discovery is a concept associated with gathering the contextual information for establishing the D2D links [6]. Moreover, cluster-based D2D networks can also be established based on the devices in proximity and the cluster head (CH) is responsible for providing the control signaling. Resource-efficient D2D network discovery mechanisms have been proposed for ensuring that the device battery is not drained due to overuse of the processing capabilities.

Device discovery mechanisms can be deterministic, where all the contextual information is available and hence utilized for discover, or probabilistic, where the algorithms utilize the tools related to stochastic systems [27]. The application of aforementioned mechanisms depends on the network requirements. For example, if low latency is required and the network needs to be made operational quickly, a probabilistic mechanism can be a suitable choice. However, if the network requires high reliability, then deterministic mechanism can be employed to gather all the contextual information and establish D2D links. Moreover, the trajectory of the devices is also a useful information that can assist the device discovery. As mentioned earlier, prediction-based algorithms have been developed that consider the mobility of devices for device discovery [28]. The device discovery mechanisms need to be dynamic so that any devices can enter or leave the network and the information is updated accordingly. Energy efficiency is one of the main concerns for D2D operation, viewing the stored battery power limitations. Therefore, research has been conducted on energy-efficient device discovery mechanisms. The frequency of device participation in discovery is managed to save the battery resources.

Device discovery mechanism can be implemented centrally, where a central authority assists the discovery process, or it can be implemented in a distributive manner where the devices conduct the discovery process. The central authority can be the BS, where the BS keeps track of all the devices in the network coverage area and performs signaling activities to aid the discovery process. Such discovery process can help the devices to save the energy as the BS performs the bulk of processing. However, in situations involving out-band D2D operation, the distributed device discovery is a desirable choice [29].

7.1.2 SWIPT for D2D

Simultaneous wireless information and power transfer (SWIPT) has been explored as a technology that can assist the devices to enhance the sustainability. The main concept behind SWIPT is that the devices decode the information and can also harvest energy from the electromagnetic signals [16]. In the network environments where the devices act as both the cellular users and the D2D users, the SWIPT can help the devices to replenish their battery storage. This can not only increase the network sustainability by increasing the battery time but also help in avoiding any delays. One of the challenges in realizing SWIPT is to gather the channel state information (CSI). The SWIPT system is designed to act dynamically based on the CSI. The idle devices harvest the energy, while the active devices transmit/receive data. There are several techniques associated with SWIPT such as time switching and power splitting. In the time switching technique, different time slots are used by the device to switch between receiving the data and energy harvesting. Power splitting is a technique where the received power is split for decoding the message and for harvesting [11].

Several advancements in the antenna models and control systems have been proposed to realize SWIPT. Employing SWIPT for D2D networks can also be helpful for managing the interference by treating any interfering signals as an energy source [8]. The energy harvested through SWIPT might not be able to cater to all the energy demands of the devices, but it can lead to an increase in the device battery time. The power control schemes designed for D2D networks are being realigned to include the SWIPT model to manage the interference as well as providing a solution for network longevity.

7.1.3 Resource Allocation

The resource allocation schemes have been devised for D2D to ensure that interference is avoided, especially in network scenarios involving both cellular and D2D networks [14]. The resource allocation schemes address both the uplink and the downlink scenarios. The allocation of resources becomes more complex when multiple D2D pairs are transmitting simultaneously, making the frequency reuse difficult to realize due to increase in interference. Optimal spectrum and power allocation schemes are important in the context of D2D networks, to maximize the network throughput and ensure efficient utilization of resources [15]. Moreover, the reliability of the network needs to be taken into consideration based on the vision of ultra-reliability for 5G networks. The interplay between cellular users and D2D users to maintain reliable transmission can only be realized due to resource allocation and interference coordination. The optimal number of D2D pairs and cellular users transmitting simultaneously can be identified using optimization tools, providing a perspective for network design. The resource allocation schemes become more important when the network design involves relays [7]. The D2D network can operate in direct mode or can utilize a relay for transmissions. The selection of optimal relay power and the mode of operation is an area of research that has drawn significant interest of the research community.

7.1.4 3GPP Standardization

In Release 12, the 3GPP recognized D2D communication as a potential contender to manage the network capacity/coverage problem, through Proximity Services (ProSe)

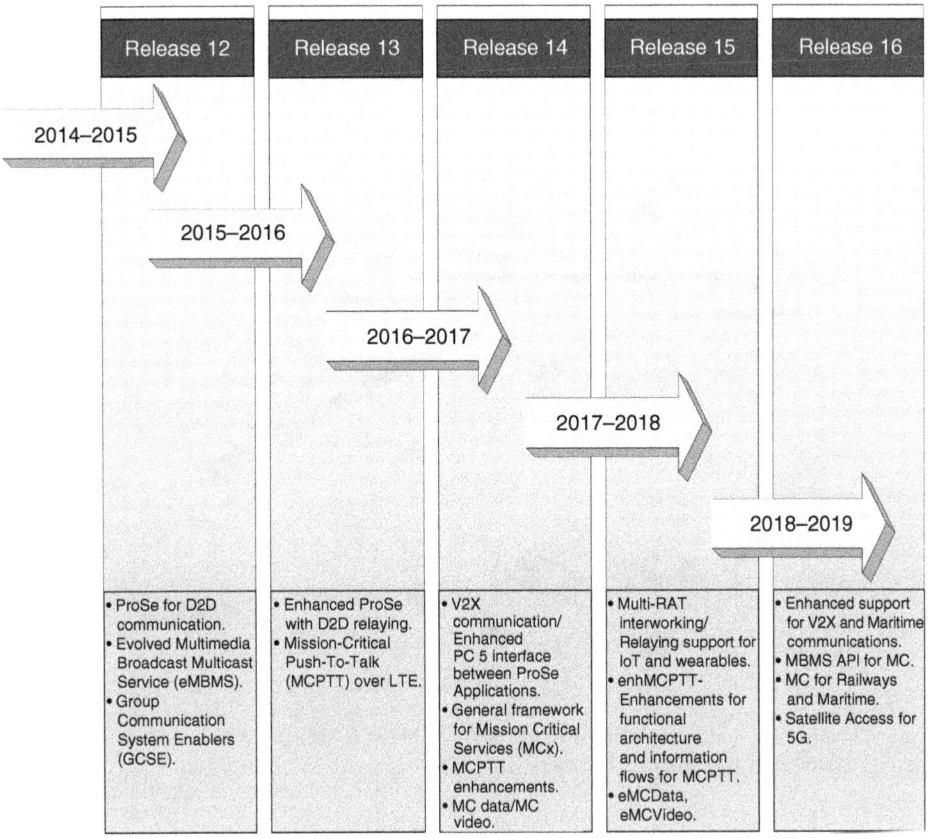

Figure 7.2 Mission critical and D2D service enhancements with 3GPP releases.

to enable D2D discovery and direct communication [17, 30]. Further enhancements in ProSe with integration to IoT and vehicle-to-everything (V2X) communications became a part of future releases as depicted in Figure 7.2.

To support such integration, such as enhancement of evolved packet core (EPC) with new features and interfaces with D2D services were addressed in [35]. Based on these results, 3GPP released another technical specification of Release 12 [36] introducing three main entities: ProSe function, ProSe Application Server, and ProSe application at the UE.

7.2 Unmanned Aerial Vehicles in D2D

UAVs [2, 3] have been utilized for assisting the traditional cellular networks by sharing the communication load. The ease of mobility of UAVs ensures their easy deployment, especially in scenarios where there is a sudden demand for network access such as hot spots. UAV-base stations (UAV-BSs) can be deployed in such scenarios and they can help in solving any bottlenecks with regards to network

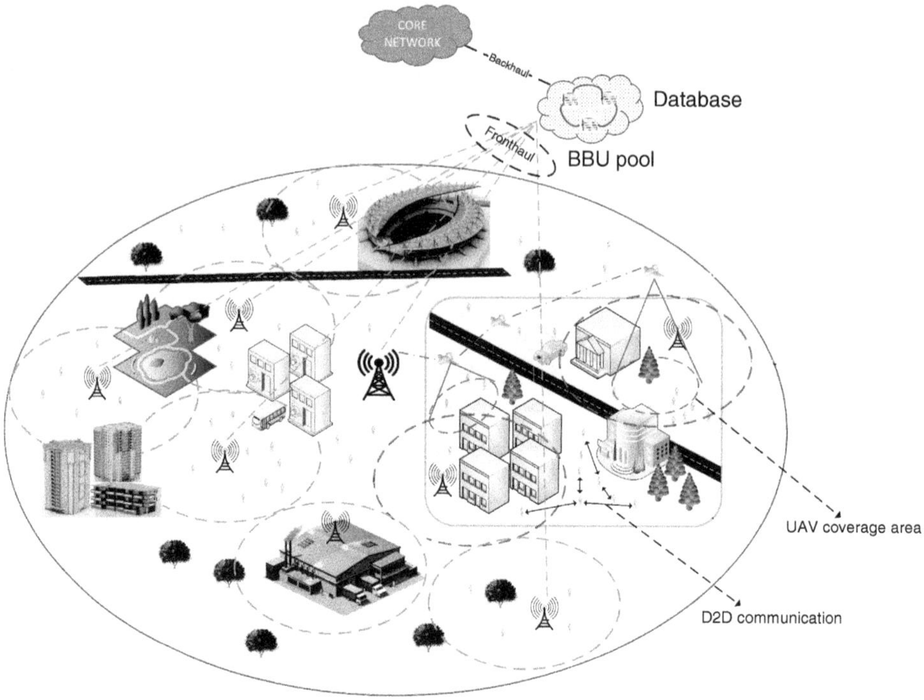

Figure 7.3 Architecture of the heterogeneous network scenario with multihop UAV and ground user equipment's D2D links.

connectivity, as shown in Figure 7.3. UAV-BSs can provide signaling for D2D networks and assist D2D networks to deliver the required content.

UAV-BSs have been proposed to assist the cellular networks to meet any new demands. The deployment of UAV BSs has drawn particular interest for providing connectivity in network hot spots. For example, a stadium might see a sudden surge in network demands due to all the people turning up for a match. In such a scenario, the cellular network can get choked as it might not be able to cater to the rise in demands. UAV-BSs can assume the role of providing connectivity in such scenarios. The ease of mobility and an on-demand deployment make the UAV-BSs a viable candidate for future networks [37]. The miniaturization of antennas and the communication modules has allowed the equipment to be mounted on the UAVs.

The UAVs can take inspiration from the D2D networks to provide an efficient backhaul mechanism. D2D networks can assist the UAV-BSs in disseminating the messages among the users in proximity, thereby reducing the transmission load at the UAV-BS. The D2D networks can also ease the pressure on the UAV-BS backhaul by sharing the transmission load [9]. The UAV-BS transmissions can be scheduled efficiently to conserve the resources. A multihop D2D network can be assisted by the UAVs to relay the transmissions to a longer distance, as shown in Figure 7.3. The cache-enabled communications by the UAVs reduce the processing overhead to demand the information from the server. The UAV-BSs can intelligently store the highly desired information in the caches so that it can be readily distributed to the D2D users. To summarize, the D2D networks can benefit from UAV-BSs to realize transmission over longer distances and assist the D2D networks to connect to the central server.

7.2.1 Key Challenges in UAV-based D2D

There are several challenges associated with the UAV-BSs. The battery constraints reduce the flight time of the UAVs which might hinder the reliability of the network. The allocation of spectrum resource for UAV-BSs is also a challenge, viewing they reuse the spectrum resources of cellular networks. Determining the backhaul network for UAV-BSs is also a research challenge, as the backhaul network needs to be designed to incur minimum delay. The research community has directed its efforts toward planning the trajectory of the UAVs [38, 39]. The localization of the devices that require the services is important for optimal deployment of the UAVs [18]. Moreover, the contextual information such as the number of devices that require the services and the channel conditions can also be utilized to identify the number of UAVs needed and optimal placement of UAVs [4]. Ensuring energy-efficient communication is also important to enhance the flight time of UAVs [40]. Several techniques have been proposed for ensuring energy-efficient communication involving UAV-BSs. The offloading of traffic from the cellular network to the UAV-BSs can provide the much needed space to the cellular networks [41]. The amount of traffic that is offloaded and the overall scheduling between UAV-BSs and the cellular networks is also an interesting area of research. The security aspects of UAV-based networks have also been explored to block any eavesdropping nodes. The secrecy performance of the UAV-based networks can help in evaluating the utility of the network in terms of trust [42].

Some ongoing and completed research projects are discussed and summarized in Table 7.1. Base Station-Aided D2D (BSA-D2D) communication in cellular networks was funded by the research Framework Program 7 (FP7) [31]. The key idea of this project was to improve the system capacity in a dense urban scenario by exploring network coding, interference alignment, joint source channel coding, regenerative storage codes, and multiple description source coding.

Table 7.1 Summary of some D2D-funded projects.

Projects	Year	Architecture
BSA-D2D [31]	2011–2012	LTE
PSS [32]	2016–2019	LTE
UAV4PSC [33]	2015–2020	5G
NICER [5]	2015–On-going	5G
DDPS [13]	2017–2019	LTE-Release 14
COUNTER-TERROR [34]	2018–2021	LTE-Release 14

Pervasive Spectrum Sharing (PSS) was a project funded by National Science Foundation (NSF) [32] for Public Safety Communications (PSNs). The focus of the project was to improve spectral efficiency in emergency scenarios by exploring game theory, public safety administration, and networking and wireless communications.

UAV-assisted heterogeneous networks For Public Safety Communications (UAV4PSCs) received funding from NSF [33]. This project proposes video streaming during disaster situations using UAVs along with cellular technologies for seamless connectivity by reducing interference between UEs, UAVs, and BS.

Networked Infrastructureless Cooperation for Emergency Response (NICER) was funded by LOEWE [5]. This project investigates on enabling cooperative communication where infrastructure to establish communication is not available. Thus, people in the affected area can maintain connectivity.

Device-to-Device System for Public safety (DDPS) was funded by National Institute of Standards and Technology (NIST) [13]. Key services of this project were to provide mission critical services with ProSe. The main goal of this project was to provide essential technologies for PSNs, and to design, implement, and test the vital DDPS components, and prototyping using Open Air Interface (OAI) platform.

Communication in Context Related to Terror Attacks (COUNTER-TERROR) has received funding from The North Atlantic Treaty Organization (NATO) Science for Peace and Security (SPS) Programme [34]. The main goal of the project is to reduce time response to law enforcement agencies by providing critical information such as the number of people trapped inside the terror zone, their identity and location, the number of terrorists, etc. To achieve this goal, technologies such as LTE-sidelink, beamforming and localization for UAVs, and multihop D2D routing for PSN in the context of terrorist attacks, are investigated.

Further, detailed summary on more projects can be found in [19].

7.2.2 Transmission over PC5 Interface for UAV-based D2D Discovery

To establish D2D communication through ProSe, several interfaces have been introduced by 3GPP, from which PC5 interface is used between two devices (for our case, UAV-to-UE and UE-to-UE). As shown in Figure 7.3, three transmission links have been considered in the multihop scenario, i.e. UE-to-UE, UAV-to-UAV, and UAV-to-UE. Here, we present the baseband simulation results for UAV-UE links for direct discovery. The Air-to-Ground (A2G) channel model and its parameters such as path loss, shadowing, and Ricean factor have been implemented from [43] for a low-altitude aerial platform. In Figures 7.4 and 7.5, PER and throughput results are shown, respectively. The simulations have been carried out for A2G transmission with UAV transmission power of 18 dBm for different modulation and coding scheme (MCS) indexes when 29 bytes of information is transmitted, as per Model A of D2D discovery [44].

The PER is minimum for lower MCS values and increases for higher values, since higher MCS provides more data but also least robust links, so the probability of losing data is higher. Furthermore, as seen in Figure 7.4, MCS 0 link provides the lowest PER for Signal-to-Noise Ratio (SNR) lower than 15 dB and MCS 3 for larger SNR

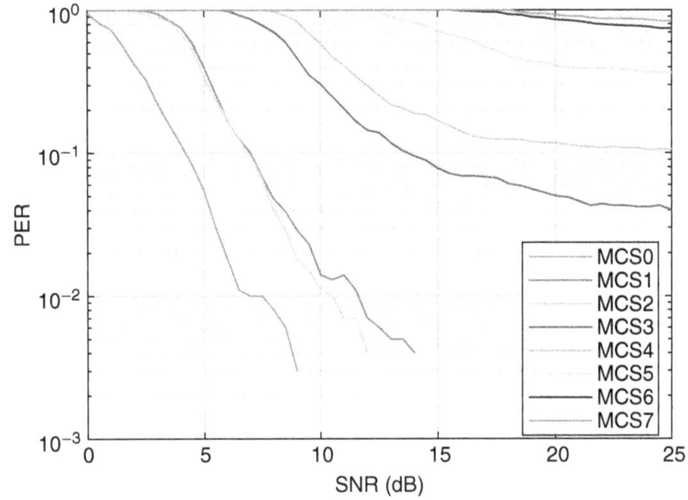

Figure 7.4 Packet error rate for UAV-UE link.

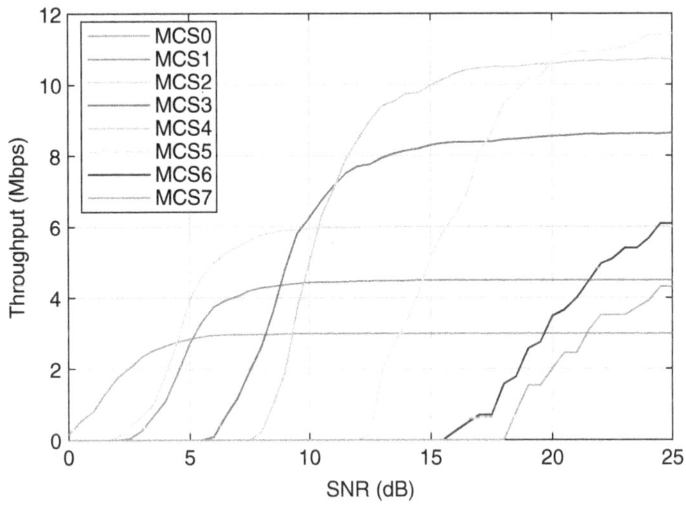

Figure 7.5 Throughput for UAV-UE link.

values. As seen in Figure 7.5, a higher throughput is achieved for higher MCS values due to more data transmission for a given range of SNR. For example, MCS 4 and 5 provide higher throughput for SNR above 10 dB while below 10 dB, MCS 2 achieves a better performance, and finally MCS 0 achieves the best throughput performance for the lowest SNR values. An adaptation mechanism could be deployed to achieve the best throughput by adaptively selecting the MCS value as a function of the SNR.

7.2.3 Interference in UAV-based D2D

In Model A of D2D discovery, transmitting UAV sends an announcement message for other monitoring UEs and UAVs to discover it. In this model, the devices can be either "announce" or "monitor" the transmission for discovery. Therefore, ideally there should not be any interference for D2D devices in Model A. However, for establishing multihop UAV communication and discovering UEs, it can be possible that multiple UAV attempts simultaneous announcements, which can cause interference on detecting UEs and other UAVs. The interference can also be caused by other heterogeneous network communication, but we do not consider it and focus on interference in UAV-UE link due to D2D transmissions.

Thus far, we use SNR as a performance metric; however, due to interference, we will address signal-to-interference and noise ratio (SINR), which can be modeled for i_{th} UE associated with u_{th} UAV as:

$$\gamma_{i,u} = \frac{P_{i,u}\, h_{i,u}}{N_0 + I_{i,k}}, \quad \forall\ i \in I \text{ and } u \in U \tag{7.1}$$

where, $I = \{i_1, i_2, \dots, i_a\}$, $U = \{u_1, u_2, \dots, u_b\}$, and $P_{i,u}$ is the transmission power of the UAV with $h_{i,u}$ as the channel gain for A2G link between UAV and UE. Further, N_0 is Additive White Gaussian Noise (AWGN) and $I_{i,k}$ is the total interference at i_{th} UE from all k_{th} interfering UAVs, which is defined as:

$$I_{i,k} = \sum_k \left[P_{i,k}\, h_{i,k}\right], \quad \forall\ k \in K \text{ and } K \subset U \tag{7.2}$$

where, $K = \{k_1, k_2, \dots, k_c\}$, $P_{i,k}$ is the transmission power of the interfering UAV with $h_{i,k}$ as the channel gain for A2G link between interfering UAV and UE.

From [12],

$$\text{PER}_{i,u} = 1 - (1 - \text{BER}_{i,u})^N \tag{7.3}$$

where BER is the bit error rate for downlink UAV to UE D2D transmission. From the results obtained in Section 7.2.2, we emphasize on BER for MCS 0, i.e. with Binary Phase Shift Keying (BPSK) modulation, due to simplicity of obtaining BER analytical expression and focusing on the effect of interference on PER.

Thus, BER expression for BPSK is given as:

$$\text{BER} = \text{Prob}\{s = +1\}\text{Prob}\{n < -1\} + \text{Prob}\{s = -1\}\text{Prob}\{n > 1\} \tag{7.4}$$

which implies that a positive bit is transmitted $s = +1$ but the received bit is negative $s = -1$ due to high noise $n < -1$. The occurrence of this event is calculated using Probability (Prob) in Eq. (7.4). Similar is true for the negative bit. This is depicted in Figure 7.6. Further, the Prob are given as:

$$\text{Prob}\{s = +1\} = \text{Prob}\{s = -1\} = \frac{1}{2} \quad \text{and}$$
$$\text{Prob}\{n < -1\} = \text{Prob}\{n > 1\} = Q\left(\frac{1}{\sigma}\right) \tag{7.5}$$

where, Q is mathematical operator, q function. From Eqs. (7.4) and (7.5),

$$\text{BER} = Q\left(\frac{1}{\sigma}\right) \tag{7.6}$$

Further, from Eqs. (7.1), (7.2), (7.6) and putting $N_0 = \sigma^2$,

$$\sigma = \sqrt{\left(\frac{P_{i,u}\, h_{i,u}}{\gamma_{i,u}} - I_{i,k}\right)} \tag{7.7}$$

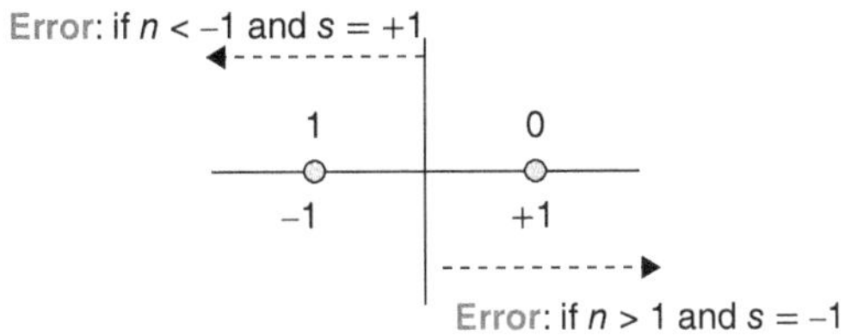

Figure 7.6 BPSK BER constellation diagram.

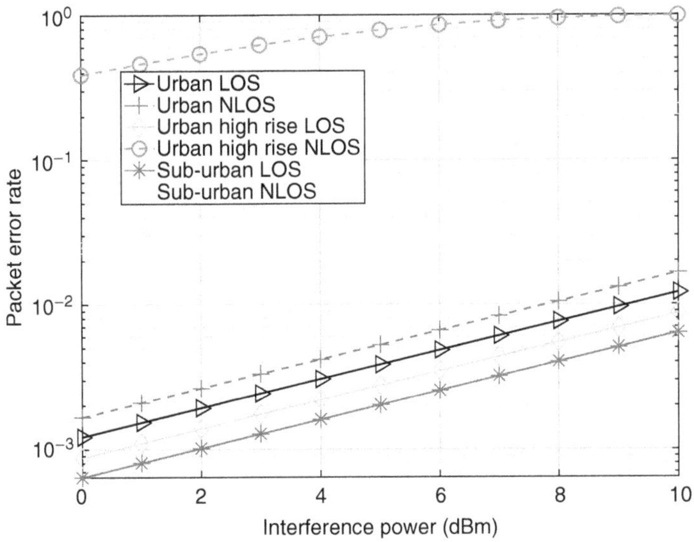

Figure 7.7 PER with varying SINR at UE for different environments.

$$\text{BER} = Q\left(\frac{\sqrt{\gamma_{i,u}}}{\sqrt{P_{i,u}\, h_{i,u} - \gamma_{i,u}\, I_{i,k}}}\right) \tag{7.8}$$

Further, PER can be evaluated using Eq. (7.3). To evaluate the impact of SINR on PER from an interfering UAV in different environment conditions, we simulate the results shown in Figures 7.7 and 7.8. In Figure 7.7, we observe a close to linear increasing behavior of PER when power of the interfering UAV (i.e. $P_{i,k}$) is increased. Also, the interfering UAV is considered to be at the same altitude, as the UAV establishing D2D connection with the UE. Further, we have shown results in different environment conditions and channel parameters as given in [4, 43]. These channel parameters were obtained from ray tracing results for UAV-UE link at 2.4 GHz in ITU-R-specified generic environments. We see for Non-Line-of-Sight (NLoS) links, higher PER is observed than Line-of-Sight (LoS) links, due to higher path-loss and shadowing effects. In Figure 7.8, we show the trend of PER when altitude of interfering UAV is increased keeping fixed the altitude of UAV connecting via D2D with UE. With increase in interfering UAV altitude, PER is expected to decrease since the interference effects are reduced. Here we kept the power $P_{i,k}$ fixed, but attenuation effects due to path loss and shadowing were changed according to [43]. For urban

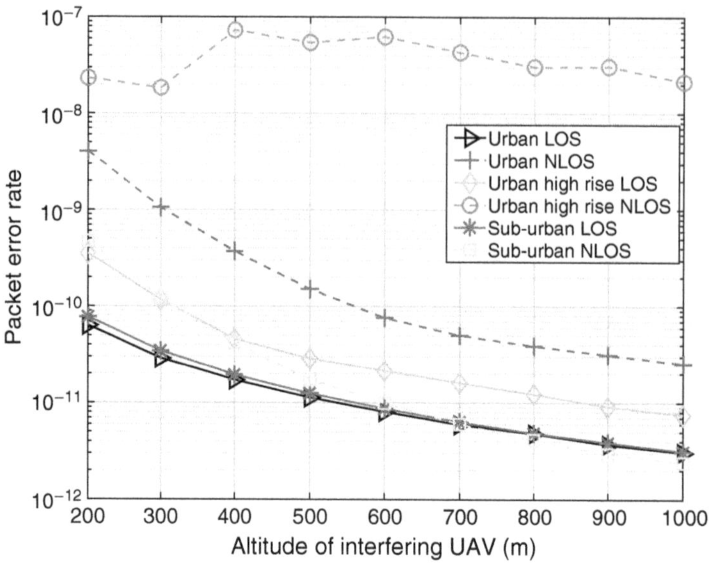

Figure 7.8 PER with varying altitude of interfering UAV for different environments.

high rise, we observe a different trend since standard deviation of shadowing is much larger due high rise building cluster in the ray tracing simulation [43].

7.3 Summary

In this chapter, we have provided details on different aspects for D2D communication, particularly for UAV-assisted D2D networks. The UAVs in a multihop fashion can relay the critical information to establish connectivity in a public safety scenario. Further, we have referred on 3GPP standardization activities and D2D research projects. To show the effectiveness of direct discovery in D2D-enabled UAV communication, we have also shown the preliminary results of PER and throughput analysis between UAV and ground UE. Further, we have addressed the impact of interference in D2D link between the UAV and the user due to another UAV establishing D2D connection simultaneously. Thus, we evaluate the change in PER with change in interfering UAV power and altitude for different environment conditions.

References

1 Bacco, M., Cassará, P., Colucci, M., and Gotta, A. (2018). Modeling reliable M2M/IoT traffic over random access satellite links in non-saturated conditions. *IEEE Journal on Selected Areas in Communications* 36 (5): 1042–1051. https://doi.org/10.1109/JSAC.2018.2832799.

2 Budhiraja, I., Kumar, N., Alazab, M. et al. (2020). Energy management scheme for wireless powered D2D users with NOMA underlaying full duplex UAV.

Proceedings of the 2nd ACM MobiCom Workshop on Drone Assisted Wireless Communications for 5G and Beyond, DroneCom '20, page 7–12, New York, NY, USA: Association for Computing Machinery. ISBN 9781450381055. https://doi.org/10.1145/3414045.3415946.

3 Chhikara, P., Tekchandani, R., Kumar, N. et al. (2020). DCNN-GA: A deep neural net architecture for navigation of UAV in indoor environment. *IEEE Internet of Things Journal* 1. https://doi.org/10.1109/JIOT.2020.3027095.

4 Cileo, D.G., Sharma, N., and Magarini, M. (2017). Coverage, capacity and interference analysis for an aerial base station in different environments. *2017 International Symposium on Wireless Communication Systems (ISWCS)*, pp. 281–286, August 2017. https://doi.org/10.1109/ISWCS.2017.8108125.

5 LOEWE-Excellent Research for Hesseas Future (2020). Networked infrastructureless cooperation for emergency response- NICER. https://www.nicer.tu-darmstadt.de/en/nicer/overview/ (accessed 08 October 2021).

6 Hashima, S., Hatano, K., Takimoto, E., and Mahmoud Mohamed, E. (2020). Neighbor discovery and selection in millimeter wave D2D networks using stochastic MAB. *IEEE Communications Letters* 24 (8): 1840–1844. https://doi.org/10.1109/LCOMM.2020.2991535.

7 Hoang, T.D., Le, L.B., and Le-Ngoc, T. (2017). Joint mode selection and resource allocation for relay-based D2D communications. *IEEE Communications Letters* 21 (2): 398–401. https://doi.org/10.1109/LCOMM.2016.2617863.

8 Huang, J., Xing, C., and Guizani, M. (2020). Power allocation for D2D communications with SWIPT. *IEEE Transactions on Wireless Communications* 19 (4): 2308–2320. https://doi.org/10.1109/TWC.2019.2963833.

9 Ji, J., Zhu, K., Niyato, D., and Wang, R. (2020). Joint trajectory design and resource allocation for secure transmission in cache-enabled UAV-relaying networks with D2D communications. *IEEE Internet of Things Journal* 1. https://doi.org/10.1109/JIOT.2020.3013647.

10 Kaleem, Z., Qadri, N.N., Duong, T.Q., and Karagiannidis, G.K. (2019). Energy-efficient device discovery in D2D cellular networks for public safety scenario. *IEEE Systems Journal* 13 (3): 2716–2719. https://doi.org/10.1109/JSYST.2019.2899922.

11 Khan, M.S., Jangsher, S., Aloqaily, M. et al. (2020). EPS-TRA: Energy efficient peer selection and time switching ratio allocation for SWIPT-enabled D2D communication. *IEEE Transactions on Sustainable Computing* 5 (3): 428–437. https://doi.org/10.1109/TSUSC.2020.2964897.

12 Khan, R., Mahtab Alam, M. (2018). Joint PHY-MAC realistic performance evaluation of body-to-body communication in IEEE 802.15.6 and SmartBan. *2018 12th International Symposium on Medical Information and Communication Technology (ISMICT)*, pp. 1–6. https://doi.org/10.1109/ISMICT.2018.8573715.

13 Communications Technology Laboratory (2017). Device-to-device system for public safety, (NIST). https://www.nist.gov/ctl/pscr/device-device-system-publicsafety (accessed 08 October 2021).

14 Lai, W., Wang, Y., Lin, H., and Li, J. (2020). Efficient resource allocation and power control for LTE-A D2D communication with pure D2D model.

IEEE Transactions on Vehicular Technology 69 (3): 3202–3216. https://doi.org/10.1109/TVT.2020.2964286.

15 Lee, J. and Lee, J.H. (2019). Performance analysis and resource allocation for cooperative D2D communication in cellular networks with multiple D2D pairs. *IEEE Communications Letters* 23 (5): 909–912. https://doi.org/10.1109/LCOMM.2019.2907252.

16 Lim, D.W. and Kang, J. (2020). Outage-optimal and suboptimal power control for D2D communications in SWIPT cellular networks with local CSI. *IEEE Wireless Communications Letters* 9 (10): 1795–1798. https://doi.org/10.1109/LWC.2020.3005428.

17 Lin, X., Andrews, J.G., Ghosh, A., and Ratasuk, R. (2014). An overview of 3GPP device-to-device proximity services. *IEEE Communications Magazine* 52 (4): 40–48. https://doi.org/10.1109/MCOM.2014.6807945.

18 Masood, A., Sharma, N., Alam, M.M. et al. (2019). Device-to-device discovery and localization assisted by UAVs in pervasive public safety networks. *Proceedings of the ACM MobiHoc Workshop on Innovative Aerial Communication Solutions for First Responders Network in Emergency Scenarios*, pp. 6–11.

19 Masood, A., Scazzoli, D., Sharma, N. et al. (2020). Surveying pervasive public safety communication technologies in the context of terrorist attacks. *Physical Communication* 41: 101109. https://doi.org/10.1016/j.phycom.2020.101109.

20 Mavromatis, A., Colman-Meixner, C., Silva, A.P. et al. (2020). A software-defined IoT device management framework for edge and cloud computing. *IEEE Internet of Things Journal* 7 (3): 1718–1735. https://doi.org/10.1109/JIOT.2019.2949629.

21 Vilgelm, M., Liñares, S.R., and Kellerer, W. (2019). Dynamic binary countdown for massive IoT random access in dense 5G networks. *IEEE Internet of Things Journal* 6 (4): 6896–6908. https://doi.org/10.1109/JIOT.2019.2912424.

22 Prerna, D., Tekchandani, R., Kumar, N., and Tanwar, S. (2020). An energy-efficient cache localization technique for D2D communication in IoT environment. *IEEE Internet of Things Journal* 1. https://doi.org/10.1109/JIOT.2020.3029168.

23 Xu, H., Xu, W., Yang, Z. et al. (2018). Pilot reuse among D2D users in D2D underlaid massive MIMO systems. *IEEE Transactions on Vehicular Technology* 67 (1): 467–482. https://doi.org/10.1109/TVT.2017.2728641.

24 Zhao, G., Chen, S., Qi, L. et al. (2019). Mobile-traffic-aware offloading for energy- and spectral-efficient large-scale D2D-enabled cellular networks. *IEEE Transactions on Wireless Communications* 18 (6): 3251–3264. https://doi.org/10.1109/TWC.2019.2912596.

25 Waqas, M., Niu, Y., Li, Y. et al. (2020). A comprehensive survey on mobility-aware D2D communications: principles, practice and challenges. *IEEE Communication Surveys and Tutorials* 22 (3): 1863–1886. https://doi.org/10.1109/COMST.2019.2923708.

26 Tang, J., Chen, G., and Coon, J.P. (2018). Joint coverage enhancement by power allocation in Poisson clustered out-of-band D2D networks. *IEEE Transactions on Vehicular Technology* 67 (12): 11537–11548. https://doi.org/10.1109/TVT.2018.2871065.

27 Nguyen, N.T., Choi, K.W., Song, L., and Han, Z. (2018). ROOMMATEs: An unsupervised indoor peer discovery approach for LTE D2D communications. *IEEE Transactions on Vehicular Technology* 67 (6): 5069–5083. https://doi.org/10.1109/TVT.2018.2832223.

28 Zhang, B., Li, Y., Jin, D., and Han, Z. (2016). Network science approach for device discovery in mobile device-to-device communications. *IEEE Transactions on Vehicular Technology* 65 (7): 5665–5679. https://doi.org/10.1109/TVT.2015.2453234.

29 Wu, H., Gao, X., Xu, S. et al. (2020). Proximate device discovery for D2D communication in LTE advanced: challenges and approaches. *IEEE Wireless Communications* 27 (4): 140–147. https://doi.org/10.1109/MWC.001.1900497.

30 3GPP TR 36.843 V12.0.1 (2014). Technical Specification Group Radio Access Network; Study on LTE Device to Device Proximity Services; Radio Aspects (Release 12).

31 FP7 Project Reference Number: 274523 (2012). Base Station (BS) Aided D2D (Device to Device) communication in cellular networks. https://cordis.europa.eu/project/id/274523 (accessed 08 October 2021).

32 NSF Ref. Nr: NSFEARS-2014-1443946 (2014). Pervasive spectrum sharing for public safety communications. https://www.nsf.gov/awardsearch/showAward?AWD_ID=1663761 (accessed 08 October 2021).

33 USARef. Nr: CNS-1453678 NSF (2015). Career: Towards broadband and UAV-assisted heterogeneous networks for public safety communication. https://sites.google.com/site/uav4psc/home (accessed 08 October 2021).

34 Milano, P. (2021). Public safety communication in context related to terror attacks, (NATO SPS). https://www.counterterror.eu/ (accessed 08 October 2021).

35 3GPP TR 23.703 V12.0.0 (2014). Technical Specification Group Services and System Aspects; Study on Architecture Enhancements to Support Proximity-Based Services (Release 12), February 2014.

36 3GPP TS 23.303 V12.1.0 (2014). Technical Specification Group Services and System Aspects; Proximity Based services (ProSe); Stage 2 (Release 12), June 2014.

37 Sharma, N., Magarini, M., Jayakody, D.N.K. et al. (2018). On-demand ultra-dense cloud drone networks: opportunities, challenges and benefits. *IEEE Communications Magazine* 56 (8): 85–91. https://doi.org/10.1109/MCOM.2018.1701001.

38 Xu, C., Yin, C., Han, W., and Wang, D. (2020). Two-UAV trajectory planning for cooperative target locating based on airborne visual tracking platform. *Electronics Letters* 56 (6): 301–303. https://doi.org/10.1049/el.2019.3577.

39 Sheikh, M.U., Riaz, M., Jameel, F. et al. (2020). Quality-aware trajectory planning of cellular connected UAVs. *Proceedings of the 2nd ACM MobiCom Workshop on Drone Assisted Wireless Communications for 5G and Beyond*, pp. 79–85.

40 Zhu, Z., Qian, L.P., Shen, J. et al. (2020). Joint optimisation of UAV grouping and energy consumption in MEC-enabled UAV communication networks. *IET Communications* 14 (16): 2723–2730. https://doi.org/10.1049/iet-com.2019.1179.

41 Azari, M.M., Geraci, G., Garcia-Rodriguez, A., and Pollin, S. (2020). Uav-to-UAV communications in cellular networks. *IEEE Transactions on Wireless Communications* 19 (9): 6130–6144. https://doi.org/10.1109/TWC.2020.3000303.

42 Ye, J., Zhang, C., Lei, H. et al. (2019). Secure UAV-to-UAV systems with spatially random UAVs. *IEEE Wireless Communications Letters* 8 (2): 564–567. https://doi.org/10.1109/LWC.2018.2879842.

43 Sharma, N., Magarini, M., Dossi, L. et al. (2018). A study of channel model parameters for aerial base stations at 2.4 GHz in different environments. *15th IEEE Annual Consumer Communications & Networking Conference (CCNC)*, pp. 1–6. IEEE.

44 3GPP TS 23.303 V16.0.0 (2020). Technical Specification Group Services and System Aspects; proximity-based services (ProSe); Stage 2 (Release 16).

8

Emergency Device-to-Device Communication: Applicability, Case Studies and Interference Mitigation

Imran Haider[1], Mohsin Raza[2], Kamran Ali[3], Muhammad Awais[2], Vishnu V. Paranthaman[4], and Muhammad Y. M. Mirza[5]

[1]*Acme Center for Research in Wireless Communications (ARWiC), Capital University of Science and Technology, Islamabad, Pakistan*
[2]*Computer Science Department, Edge Hill University, Ormskirk, UK*
[3]*Computer Science Department, Middlesex University, London, UK*
[4]*RDS Global Ltd, Derby, UK*
[5]*Lahore Business School, University of Lahore, Lahore, Pakistan*

Communications is a fundamental element of disaster management service and rescue response. While the cellular communication is one of the most widely available communication infrastructures and can support communications from remote areas, however, its fragility cannot be denied. The stand-alone and centralized approach in cellular networks makes it more prone to failure in disasters and natural catastrophes. The communication in cellular networks highly relies on centralized entities such as base station subsystem (BSS), mobile switching center (MSC), etc. The BSS and MSC have higher chances of failure in disasters, thus resulting in communication failure in the entire region. Device-to-device (D2D) communications on the other hand is an infrastructure-less communication paradigm, which proves to be far more resilient in natural disasters. While D2D cannot supersede the bandwidth and data rate of cellular networks in normal circumstances, its ad hoc nature, resilience, and scalability attributes make it a suitable alternate for re-establishing communication in disaster-affected and communications-outage area. This chapter focuses on the primary attributes of D2D communications for emergency and rescue services. The chapter highlights the salient features of D2D communications infrastructure which makes it an essential paradigm to support critical and emergency communication services. The key application areas and potential uses of D2D communications are presented. The applicability of D2D communications in different scenarios is also highlighted. The discussion also analyses the suitability of D2D communication in diverse scenarios. Finally, the chapter presents the case studies to highlight the potential of D2D communication infrastructure in natural disasters and catastrophes and discusses suitable interference mitigation techniques.

Interference Mitigation in Device-to-Device Communications, First Edition.
Edited by Masood Ur Rehman, Ghazanfar Ali Safdar, and Mohammad Asad Rehman Chaudhry.

8.1 Emergency D2D Communication

Climate change and global warming are serving as the driving forces for increasing natural calamities. Disasters are witnessed more frequently in the past few years. These natural disasters can severely affect people's life and can cause significant damage to human life, properties, and socioeconomic cycle. Natural disasters are mainly characterized in three categories [1]: meteorological, hydrological hazards, and geological disasters. These three mainly compose and categorize all major natural calamities, including tornados, thunderstorms, winter storms, hurricanes, wildfire, coastal flooding, flash floods, river floods storms surges, tsunamis, earthquakes, land sliding, volcanic eruptions, and avalanches. In all natural calamities, while the impact on livelihood of the victims, socioeconomic circle, and cycle of life is severely affected, the impact could be even worse with poorly managed post-disaster rescue and support services. In many cases, the difference between life and death merely equates to timely action and the effectiveness of the response teams. While the disasters may affect a large area, there could be certain vulnerable and time-sensitive regions/grids in need of immediate attention. In such circumstances, the communications play a very important role and can be very instrumental in making crucial decisions and allocating resources in real time. Unfortunately, the more critical regions also suffer from failure in communications infrastructure and any radio, tele-broadcasts, and cellular base stations (Bss) in the area are rendered useless [2]. Failure in power supply also limit the battery life of handheld devices and suitable communications and guidance to the vulnerable survivors within the affected area need response in a very short period of time [3]. The destruction of bridges and road infrastructure also limit accessibility to heavy transportable BSs. There are some other alternates including drones [4], long-range emergency broadcasts, but these are limited by their scope, suitability to the environment, and applicability. Besides, in such cases, a more robust and effective means of communication reestablishment mechanisms are required.

D2D communications offer a robust and scalable solution in disasters [5–7]. While D2D communications generally refers to the communications of end devices between each other and in some cases is considered a communication scenario where only two devices are involved [8]. However, when discussed in disaster scenario, the D2D communications offer effective mechanism to relay information over long distances with ad hoc networks formed on demand. In this section, an overview of D2D communications is covered where basic question regarding D2D communications, its potential to be widely adopted, and suitability in diverse scenarios is discussed. It also discusses possible implications of technology solutions of their transformation to be used in the event of disasters.

The significant rise in the mobile devices, use of sensory networks, internet of things (IoT), communication-enabled devices, vehicular networks, and long-range communication systems offer a plethora of communication entities which possess the capabilities of D2D communications [9]. Most of the communication devices are capable to handle large amounts of data. Some such devices like cell phones and 802.11p-enabled vehicular networks possess enough power to serve as a data sink or cluster head to keep the entire network running. While the high-throughput capabilities of cellular devices make them suitable choice in data-centric communication, it still lacks in battery life. On the other hand, the vehicular networks offer enough battery to survive prolonged operations until conventional communication systems are up and running [10].

The conventional communications systems have relatively low resilience and in the event of any natural calamity, the cellular networks, tele-broadcasts, even broadband stops working. The networks are rendered useless in the disasters and cellular and other communication devices, though still working, become inoperable. In an era where a single person owns multiple devices with wireless capabilities and IoT-enabled devices increasing at an exponential rate, D2D communications become an obvious choice to address the communication challenges during disasters. The D2D communications allow all these devices to form ad hoc networks where different responsibilities are allocated to individual devices to manage the communications and data routing between the D2D mesh devices. Usually, polling-based decisioning is followed where cluster heads are selected to communicate information from one section of the network to the next. The network size in D2D communications is highly scalable and can be formed in relatively less time which gives real edge on other communication solutions proposed over time. In this way, the communication of an individual device (for instance a cellular mobile phone) can be extended quite significantly where multi-hop relaying is implemented. Multi-hop information relaying also allows the communication-outage area to be connected with the backhaul network. This not only provides access to emergency broadcasts for the trapped survivors, but also a prioritized communication can be established to assist rescue workers and disaster support teams. In addition, the data from the sensory elements present and functional within the disaster-affected area can be relayed through established D2D communications to the disaster control center. These data can help in forming risk maps of the disaster-affected area which can be used to form action plans and support and rescue decisions in real time.

In disasters, network load significantly increases and even with fully functional networks, the system is overloaded and overhead becomes too significant. This results in the network congestion and causes severe communication delays for both emergency and nonemergency communications. The D2D communications in such scenarios can help improve the spectrum utilization by enabling direct communications between two devices without involving the BS. The cellular framework also facilitates such services. The D2D communications serve as an ideal candidate for such scenarios and can offer significant improvement in spectrum utilization, congestion, queuing, latency, and quality of service (QoS).

The recent research trends in D2D communications and exploitation of D2D features in diverse applications have resulted in significant improvements in the proposals and D2D communication frameworks. In the past, D2D communications were mainly viewed as a solution for small-scale networks and its use was primarily investigated for unlicensed spectrum. However, in the recent years, D2D communications have taken precedence over several other research themes.

The D2D communications have become an essential part of many standards. LTE release 12 also introduced support for D2D communications [11]. This has opened new venues for research in D2D communications inspired by earlier and ongoing research [9, 12, 13].

8.2 Approaches for Efficient Emergency D2D Communication

Communication failures are quite common issues in disastrous situations, and it is vital to establish communication in adversely affected areas. Therefore, reliable and effective communications in such pre- and post-disastrous situations between the affected and the responders are essential to exchange critical and emergency information through voice or text messaging services. The D2D communications, in this regard, provides an effective and reliable alternative to establish emergency communications due to its capabilities of achieving sufficient throughput in short-distance links and the advantage of providing direct communications along with the capacity to handle multiple users simultaneously through ad hoc network formation and clustering [14].

In the recent years, researchers have made significant developments in D2D communications and incorporated a variety of ways to compensate for communication failures in pre-disaster as well as post-disaster scenarios. There are variety of ways D2D communication can play a vital role to establish effective communication in network dysfunctional areas during a disaster. It can leverage the communication beyond the communication active area through multi-hop relaying. An example scenario of D2D communications is presented in Figure 8.1. The D2D communications have the potential to establish communications in areas where no means of communication, thus extending the network coverage through multi-hop relaying. This allows D2D-based ad hoc networks to form communication between healthy and dysfunctional/communication-outage regions, interconnecting the two. The D2D communications also benefit from short-distance multi-hop communication which extends the battery life of the devices. This is especially important in disaster-affected areas with power. The adopted multi-hop communications with some energy-harvesting techniques can extend the network lifetime significantly, thus allowing emergency messages and information exchange active over longer periods of time. Another energy resource where D2D can tap in is the vehicular networks, where vehicles can be used to communicate information over longer distances, thus conserving the battery of the cellular mobile phones. The access to use both licensed/unlicensed bands also enhances the bandwidth efficiency of the D2D communications.

Over the years, many works have investigated D2D communications and presented new proposals to improve the performance and efficiency of D2D communications. Some of these works provide frameworks for information exchange and emergency assistance in areas with partial or complete communication failure. Table 8.1 lists some of the research works which utilized D2D communications to assist in disaster management, whereas details of these works are listed as follows.

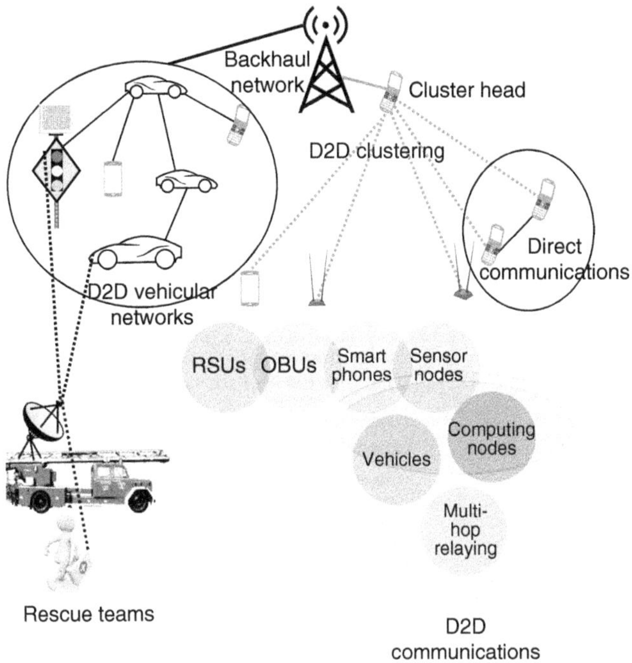

Figure 8.1 D2D communication scenario.

More recently, Raza et al. [1] developed a novel D2D-based framework which exploits clustering scheme to establish communication in natural disasters. The work also proposed optimization scheme to enhance the throughput and the localization performance of end users. The results are thoroughly validated, and the findings suggest that the proposed ad hoc clustering scheme and optimization scheme have outperformed state of the art and can effectively establish communications with improved throughput in newly developed clusters with better localization capabilities.

Ali et al. [21] proposed a novel energy-harvesting mechanism to extend the energy of D2D networks by exploiting the energy from the radio frequency networks wirelessly via BSs and accumulating clustering method and D2D into the wireless communication networks. The proposed D2D communications method is intended to support the communication services in emergency situations when the cellular network is partially damaged. The findings of this work are thoroughly validated and suggested that the proposed energy-harvesting model integrated with D2D has effectively enhanced the coverage area and coverage quality and improved the battery life of the system. A disaster response network is proposed in [15] where D2D communications were used to optimize cooperative beamforming vector and enhancing the relay selection through solving a mixed-integer nonlinear problem. A multi-hop strategy was adopted via clustering topology to establish communication between the transmitter and the receiver for the relaying network. The node section for the relay is accomplished by analyzing the power received and power leaked at the primary user's destination ends.

Table 8.1 Studies utilizing D2D communications in disaster management.

Author	D2D aspects	Role	Disaster category/ pre- or post-
Raza et al. [1]	Single- and multi-hop communications, throughout optimization	New cluster formation to improve communication between EU end users and cellular networks	Natural disasters/post disaster
Ali et al. [14]	Coverage, energy efficiency, and cluster formation	Energy-harvesting of cellular networks	Natural disasters/post disaster
Moghaddam et al. [15]	Multi-hop relay networks, hierarchical clustering	Supporting long-distance communication through relay selection and cooperative BEAMFORMING	Natural disasters/ post-disaster
Zhao et al. [16]	Providing multi-hop communications	Coupling UAV with D2D to extend coverage	Natural disasters/ post-disaster
Al-Hourani et al. [17]	Multihop D2D relaying	Extending coverage from health to damage network cells	Natural disasters/ post-disaster
Nishiyama et al. [18]	Multi-hop communication	Utilizing smartphones as relay nodes to exchange emergency messages	Tsunamic, hurricanes/post-disaster
Ali et Al. [19]	Energy efficiency, multi-hop communication	D2D pair communications established with other EUs	Post-disaster
Ahmad et al. [7]	Energy efficiency, energy optimization	Extending EUs device battery life up to 11 hours by introducing survival on sharing	Natural disasters
Thomas and Raja [20]	Energy efficiency of nodes, extending network survival time, dynamic clustering	Localizing isolated nodes to utilize for emergency services	Natural disasters/ pre-disaster

In the aforementioned studies, D2D communications are used to establish emergency communication alone by incorporating several improvements in coverage. Researchers have integrated D2D with other networking methodologies to establish emergency communication. Zhao et al. [16] developed unmanned aerial vehicle (UAV)-based network to establish communication in disasters where UAV acted as a remote BS, and D2D communications' multi-hop properties were exploited to extend the coverage area effectively. A stochastic geometry-based solution was provided for D2D communication in [17] which extended the coverage in disaster-affected areas. The aim was to elevate the network coverage from healthy network regions to defected network regions via D2D where a buffer-based analysis was performed to observe how much damage can be absorbed by the network while maintaining the normal service. The proposed analytical method used the real network parameters for model validation and to observe the effectiveness of proposed solution.

Nishiyama et al. [18] proposed a smartphone-based solution to establish multi-hop D2D communication during natural disaster where smartphones acted as relay nodes. The study successfully developed and tested prototype by utilizing users' smartphones devices as relay nodes and means of communication in field experiments. The findings showed that the smartphones in the form of relay nodes can provide effective communications regardless of the network operators and can disseminate emergency information to evacuation centers and affected and emergency services in dysfunctional network regions.

Most of the aforementioned studies utilizing D2D communication have aimed for improved network substantially, network coverage, and energy optimization, while some have incorporated energy harvesting. The aim was to increase the lifetime of the network and to extend the convergence to regions where the cellular networks are partially damaged or fully damaged. However, an important aspect to consider is that nearly all of the studies have incorporated simulation-based tools or analytical reasoning to support the research findings. An important missing aspect is the near real-life validation or testing of such scenarios in more challenging and adverse environments that are much closer to the situations faced during natural disasters.

8.3 Emergency D2D Communication: Case Studies

In public safety or emergency situations, feasibility of D2D proximity services are studied with the purpose of recognizing use cases and potential requirement for discovery and communications between user equipments (UEs) that are in proximity, including network operator control or direct [14].

i. *In-coverage*: This scenario indicates that all the considered UEs are within eNB coverage (in-coverage UEs) to receive services/signals from an evolved Node B (eNB).
ii. *Partial-coverage*: This scenario indicates that some UEs are within eNB coverage, while other UEs are outside eNB coverage.
iii. *Out-of-coverage*: This scenario indicates that all the considered UEs are outside eNB coverage (out-of-coverage UEs) and cannot receive services/signals from an eNB.

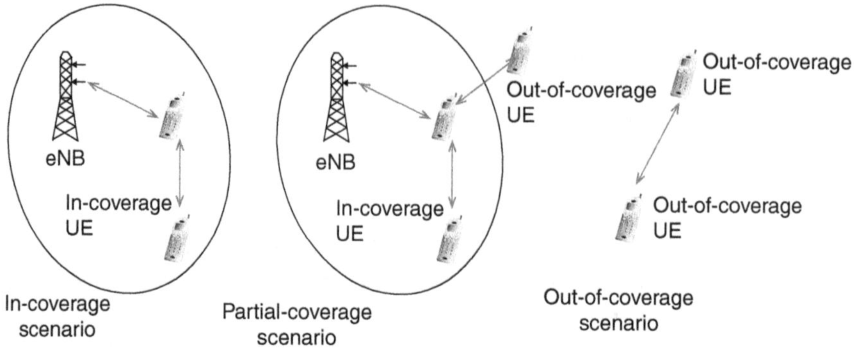

Figure 8.2 Features of D2D communication.

D2D communication is useful not only for local communication but also for communication between a BS and out-of-coverage UE. This facility can be enabled by D2D-based relay [14], and an out-of-coverage UE can communicate with a neighbor UE via D2D communication; the later UE becomes a relay to the BS as shown in Figure 8.2. These features were standardized in 3GPP Release 13 [21, 22]. D2D communications are significantly considered in the works to deliver substantial benefits, including important developments in spectrum efficiency, transmission data rates using higher-order modulation and coding schemes, power saving, and latency performance. Nevertheless, achieving above-engineering potentials are not the key design goals of D2D ProSe. Instead, a new feature of supporting communications with or without coordination from the network has been created, which leads to two components of D2D ProSe:

i. Through network-assisted discovery, UEs in close physical proximity are able to discover the existence of each other.
ii. Direct communications among two UEs can happen with or without control from the network [23].

We study D2D communication devices and the practice of multi-hop communications in public safety situations. From a technical perspective, manipulating the nature of proximity can deliver numerous benefits in disaster situations. These reduces end-to-end delays, saves energy, improves radio resource utilization, offloads cellular traffic, alleviates congestion, and enables extension for cellular range via UE-to-UE relaying We considered one scenario of multi-hop D2D communication with two modes.

i. *Mode-1*: Four predefined paths in which UEs intend to exchange information with each other using multi-hop D2D and extend communication from overage area to nonfunctional area.
ii. *Mode-2*: Using multi-hop D2D communication to cover nonfunctional area and exchange information.

Several metrics are analyzed in this work, i.e. number of hops and distance between them, energy and capacity efficiency, extendable coverage area, and

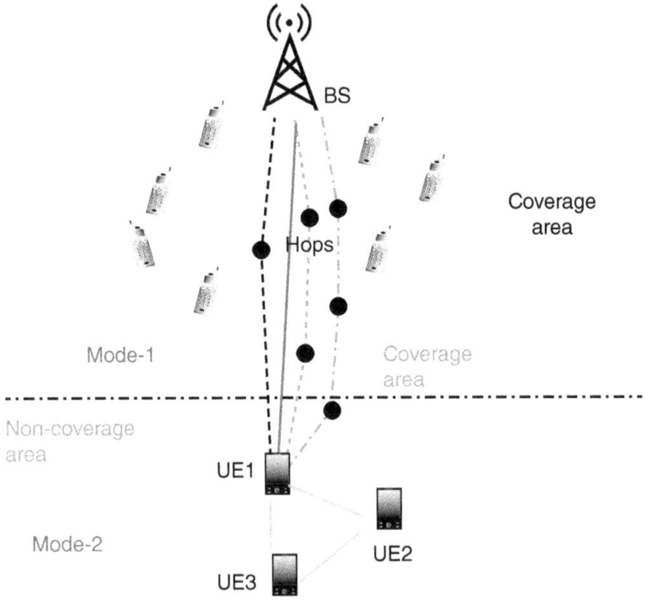

Figure 8.3 System scenario for D2D.

support communication among UEs by finding quick and shortest path for communication. Two algorithms are implemented in this work (shortest path routing and interference aware routing) and on the bases of matrices, only shortest path routing in our scenario. For communication purpose, D2D scenario is considered underlying a cellular network, where UE1 intends to communicate with active BS nearer to its position. As shown in Figure 8.3, in mode-1, UE1 is located in noncoverage area and try to connect with the BS1 directly or by using different UEs as relay. Furthermore, for simplicity, four different pathways have been predefined: path-1 (direct communication with BS [0-hop]) not possible because of the distance, path-2 (communication with BS using one UE as relay [1-hop]), path-3 (communication with BS using two UEs as relay [2-hops]), and path-4 (communication with BS using three UEs as relay [3-hops]).

Energy efficiency (E_e) and spectral efficiency S_e have developed two of the dynamic requirements in the design of future public safety communications architecture scheme. Due to the expected boundaries in energy and spectral resources during emergency situations, D2D communication should be utilized in order to increase the cellular spectrum efficiency. This can be done and control with proper interference management and resource allocation. As shown in Figure 8.3, every communication path within a route is independent and follows different time slots.

According to scenario mode-2, distance between UE1 and UE2 is d_{UE} and it is known that UE3 is in the middle of both $d_{\mathrm{UE}}/2$. The transmission power of UE_t where $t = 1,2,3$ is the set of P_{UE} and channel gain C_g between UE_t and BS. In this mode, coverage in uncovered area is extended, and UE1 and UE2 will exchange

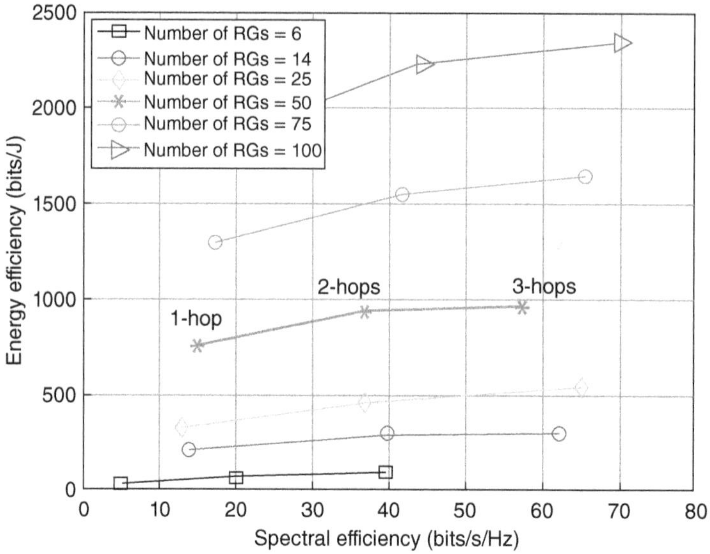

Figure 8.4 Energy efficiency of the multi-hop D2D transmissions.

information via UE3 within two time slots. In the first time slot, both UE1 and UE2 transmit to UE3 at the same time.

The simulation results are computed to show the performance of an efficient resource management using multi-hop D2D communication. As BS knows every channel in the field, it can easily determine which node can perform best in the area. In order to select relay shortest path toward BS, shortest- path-routing (SPR) algorithm is required. Energy efficiency as a function of spectral efficiency for variations in resource groups (RGs) are presented in Figure 8.4. Simulation results show that increasing the RGs will benefit both energy efficiency and spectral efficiency of the multi-hop D2D transmissions. Energy efficiency rises for the same UE transmission power. This is due to the fact that UE acts as a relay which influences the system capacity to improve due to the shorter communication path, which results better channel conditions. Also, these results show that the use of shorter D2D links while increasing the number of hops guarantee or even improve the QoS. The spectral efficiency increases when using higher number of RGs for the same number of hops. This is due to the fact that RGs increase is associated with bandwidth increase. However, when using higher number of hops, we also notice that energy efficiency is bounded by a certain limit. This limitation is due to higher aggregate interference effects in multi-hops communications.

8.4 Interference Mitigation in Emergency D2D Communication

Freedom of direct communication among D2D devices or vehicles within certain area has many benefits for network Base stations (eNB for LTE/4G) and other communicating devices. The benefits are like, reduced overhead of network traffic, more users served due to reuse of radio resources, fast link establishment, and favorable channel conditions which are quite helpful in rapid deployment of communication infrastructure in case of natural or manmade disasters. All these benefits of D2D communications are elaborated in detail in the earlier sections.

However, devices communicating directly with each other have minimal coordination with eNB, so the instantaneous communication parameters are not known to the network, such as selected frequency band and its allocation scheme, transmission power level, requirement of multiplexing or diversity gain due to mobility and their location, etc. This can negatively affect service for demanding cellular users within the eNB service area due to cross-tier interference in communication link. Another scenario may be direct link between the devices which can be established with zero coordination with eNB at no network coverage regions. However, eNB, cellular or D2D users might sense multiple signals at receiver antenna which are transmitted through users from farther distance, but same frequency band or resource block might be allocated in the network for Uplink/Downlink. There might also be some nearby cell transmitting at a tightly adjacent frequency band. These multiple signals decrease the signal-to-interference and noise ratio (SINR) at receiver and cause disruption in the communication link of users. The signal which disrupts cellular user due to sharing of common radio resource is called cross-tier interference.

Dedicated short range communication (DSRC) is 802.11p-based vehicle-to-vehicle (V2V) communication protocol (5.84 GHz) for communication between automotive vehicles without involvement of cellular network. It is currently implemented in the United States for short-to-medium-range communication, and it is prone to experience undesired interferences from cellular network communications [24].

Third-Generation Partnership Project (3GPP) Release 16 for 5G New Radio (NR) has recently announced its support for vehicle-to-everything (V2X) systems including for V2V communication links to ensure reliable channel latency with ultra-high bandwidth (at carrier frequencies 27 GHz, 79 GHz, or 300 GHz) with excellent mobility management. However, more undesired interference scenarios are noticed with increase in carrier frequencies (smaller wavelengths) toward spectrum of 30–300 GHz. The analysis of interference level and their mitigation schemes are being worked out [22].

Several interference mitigation schemes are highlighted in this section for most of the D2D failure scenarios discussed above.

8.4.1 Radiated Power Management

Power of the communication signals radiated through the antennas of eNB and D2D users must be regulated after every transmission interval to remain below the defined limits (dBm), because D2D and network users mapped at the same resource blocks for reuse of frequency bands must not interfere with each other [25]. Open-loop transmissions are likely to cause interference for cellular users because their SINRs are prone to degradation from nearby D2D users. Power control helps in better utilization of limited spectrum without disruption. The uplink power constraint level largely depends on the distance between communicating D2D devices and their displacement from the closest eNB station. If eNB is far away, then devices can transmit at relatively high power for reliable link if their mutual distance is large. This is helpful specially to combat fast channel fading effects for the moving devices, like in V2V communication. However, if communicating vehicles are in close proximity of network eNB then their power level must not exceed the constraint levels. Similarly, downlink signal power from eNB and cellular users should not degrade SINR of vehicles at relatively larger distances. Power control scheme can mitigate the effects of cross-tier interference, but it is only feasible if communicating devices are mutually close to each other or they are at relatively long distance from the nearby eNB [26].

8.4.2 Frequency Allocation

Intelligent allocation of frequency bands to D2D and cellular users is a productive approach in interference mitigation. Signal power at eNB is checked for each transmission interval to estimate the distance between eNB and D2D nodes or vehicles. Assignment of radio frequency bands for reuse in direct links between devices is performed for those frequencies which are already being used for Uplink to eNB or Downlink from eNB. Dynamic allocation of frequency bands after regular time interval improves network effectiveness for cellular and D2D devices at the cost of high signaling overhead for eNB. Several other location-based and distance-based spectrum allocation schemes are also suggested to mitigate the interference along with robust frequency reuse mechanism and ease of signaling overhead for eNB [26].

8.4.2.1 Hybrid Schemes for Power Control and Intelligent Frequency Allocation

LTE network supports independent communication links among devices while serving other cellular users by utilizing modern schemes of dynamic signal power regulation. D2D and network users managed through suitable spectrum allocation techniques to yield benefits of both schemes and achieve optimal network performance. This hybrid approach can be practically implemented by excessive signaling from eNB. Initially, assignment of radio resources is performed by eNB to the cellular users, and then D2D users are allocated with the suitable frequency band. However, if more requests are coming for new D2D links then frequency bands of cellular users are reutilized for new D2D links with eNB-governed dynamic control of signal power [26, 27].

Distance-based allocation: Distance between the D2D users and closest nearby cellular users is periodically estimated at eNB and it is analyzed to allocate similar Uplink frequency bands only if they are not in close proximity of safe distance to avoid interference. Moreover, transmission power levels of D2D devices are regulated through eNB to avoid interference among communicating D2D and cellular devices [28].

Location-based allocation: In this approach, cellular users and D2D devices do not share similar frequency bands within area specified by the eNB. Therefore, no load of signaling overhead for eNB to manage the spectrum reuse and interference is also avoided which improves the network performance at the cost of less users can be served [29].

8.4.3 Time Division Multiplexing (TDM)

Allocation of separate time slots for cellular network users and D2D users is the easiest method to mitigate cross-tier interference between the both types of users. However, radio resources utilization would be inefficient. Moreover, interference among the D2D communication pairs cannot be avoided through this method. Therefore, this scheme can be used along with another interference mitigation scheme to achieve overall better network performance [30].

8.4.4 Adjacent Channel Interference Cancellation in DSRC

Radio links between vehicles which utilizes D2D communications are integral part of overall V2X network. The protocols for DSRC links (5.8 GHz) among moving vehicles are defined in the IEEE standard 802.11p which is applicable to both DSRC and Intelligent Transportation-G5 (IT-G5) which are foundation standards implemented in the United States and Europe for vehicular communications.

A novel interference cancelation scheme for mitigation of mutual interference on electronic toll collection Road Side Unit operating in DSRC frequency 5.8 GHz and ITS network operating at 5.9 GHz [24]. Schematic of Interference cancelation approach is illustrated in Figure 8.5.

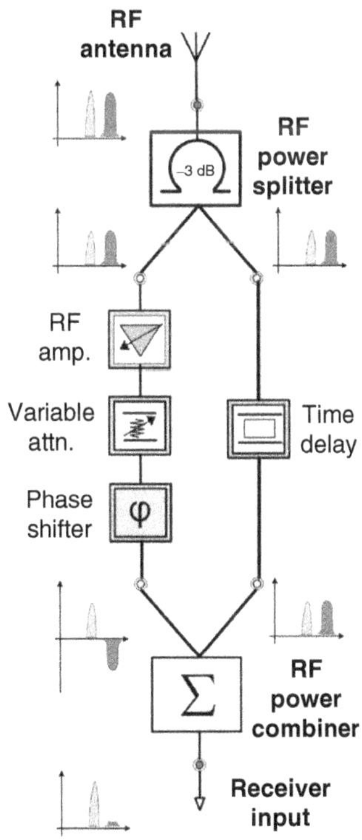

Figure 8.5 Analog interference mitigation scheme for adjacent channel problem. Source: Adapted from [24].

Table 8.2 Parameters of RSU unit and ITU signal.

Parameter description	Value
RSU EIRP	33 dBm
ITS EIRP	33 dBm
Interfering signal strength	−60 to −20 dBm
Multiplexing scheme	OFDM
Bandwidth	10 MHz
Worst SIR	−50 dB
Distance (RSU OBU)	07 m
RSU antenna gain G_t	12 dBi
DSRC signal strength	90–80 dBm

The parameters of DSRC and ITS signals are stated in Table 8.2.

The mechanism of interference cancelation consists of processing blocks of power dividers at both ends, variable gain amplifier, attenuator, and phase shifter. Signals coupled at DSRC receiver antenna which have desired power at 5.8 GHz band and interference power at 5.9 GHz band. These signals move along the two paths, one path is just delayed and used for final cancelation, while the other path is processed for magnitude stabilization and phase inversion of 5.9 GHz power so that the inverted imitation at the second divider is canceled while keeping integrity of the desired 5.8 GHz signal. The process is depicted in Figure 8.5, and the scheme achieves signal-to-interference ratio (SIR) improvement of up to 25 dB within 10 MHz bandwidth [24].

8.4.5 Interference Mitigation through Multiple Antennas (MIMO)

Transceivers used for communications in 4G and 5G cellular networks are very much composed of multiple transmitter and receiver antennas also called multiple input multiple output (MIMO) systems. Multi-antenna-based transmission schemes like beam steering towards the desired directions are quite promising for better performance for interference-free D2D exchange of information along with existing network services to cellular users [31].

Three configurations of uniform planar antenna arrays are considered to independently estimate the impact of interference as average SINR which is function of intergroup distance as illustrated in Figure 8.6. The curves clearly show that directional communication through antenna arrays mitigate effects of interference and give better SINR gain of around 10 dB each if antenna configurations are changed from 1×1 to 2×2 and 4×4, respectively [32].

8.4.5.1 Beam Steering in 3GPP 5G NR Supported Vehicular Systems

V2V the latest 5G standard defined by 3GPP Rel. 14NR is which supports vehicular systems including Vehicle to Vehicle (V2V) Communications at mmWave frequency

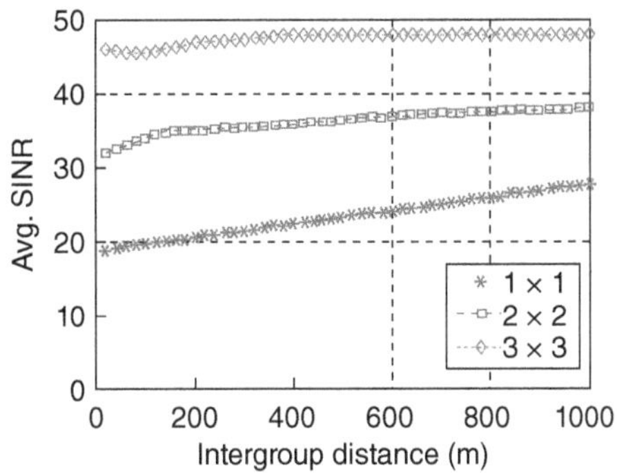

Figure 8.6 Average SINR versus distance between two groups of vehicles for three $N \times M$ antenna configurations.

spectrum (30 GHz to 300 GHz). The requirements of smart automotive networks for high bandwidth, low latency, and reliable link in mobility are met by using carrier frequencies beyond 3 GHz in cellular networks. However, characteristics of wireless channels at mmWave range becomes quite challenging due to the interference caused by simultaneous transmissions from nearby vehicles, reflections or scattering from rough surfaces, and metallic bodies of roadside vehicles due to shorter wavelengths. Therefore, multiple antennas of transceivers at 5G base stations and automotive vehicles enables steering of radiation beams towards the intended mobile user/base station, separating users in space and allow interference-free directional communication. Adaptive beam tracking and beam steering towards the positions of users in open environment like highways, etc. results in better performance in terms of increased average SINR. An open vehicular environment is simulated in which two groups, each with two vehicles are moving towards the same direction on different lanes of a highway. All the vehicles have a Uniform Array of $N \times M$ Planar Antenna elements which enables directional beams at carrier frequency of $f_c = 28$ GHz, 100 MHz bandwidth [32]. All the considered simulation parameters are listed in Table 8.3.

Table 8.3 Parameters for intergroup V2V communication.

Parameter description	Value
Constant vehicle speed	20 m/s
Distance between two vehicles	40 m
Intergroup distance	80–600 m
Carrier frequency	28 GHz
Bandwidth	100 MHz
Packet generation rate	100 ms
Statical distribution for packet generation	Exponential distribution
Uniform planar array ($N \times M$)	1×1 2×2 4×4

8.5 Summary

This chapter has highlighted the primary attributes of D2D communications for emergency and rescue services. Key application areas and potential uses of D2D communication infrastructure in natural disasters and catastrophes have been discussed by taking up on different practical case studies. D2D offers a convenient and effective means to facilitate the emergency services in the event of damaged/dysfunctional conventional communication infrastructure. Interference, however, is a burning issue and can hamper the rescue services greatly. Usability of various interference mitigation techniques including power control, frequency allocation, time division multiplexing (TDM), adjacent channel interference cancelation, and MIMO antennas in these scenarios has also been considered.

References

1 Raza, M., Awais, M., Ali, K. et al. (2020). Establishing effective communications in disaster affected areas and artificial intelligence based detection using social media platform. *Future Generation Computer Systems* 112: 1057–1069.

2 Keller, E. and DeVecchio, D. (2015). *Natural Hazards: Earth's Processes As Hazards, Disasters, and Catastrophes.* Pearson Higher Education AU.

3 Raza, M., Hussain, S., Le-Minh, H., and Aslam, N. (2020). Novel MAC layer proposal for URLLC in industrial wireless sensor networks. *ZTE Communications* 15 (S1): 50–59.

4 Ali, K., Nguyen, H.X., Vien, Q.-T. et al. (2020). Deployment of drone-based small cells for public safety communication system. *IEEE Systems Journal* 14 (2): 2882–2891.

5 Rihan, M., Selim, M.M., Xu, C., and Huang, L. (2019). D2D communication underlaying UAV on multiple bands in disaster area: stochastic geometry analysis. *IEEE Access* 7: 156646–156658.

6 Liu, X., Li, Z., Zhao, N. et al. (2018). Transceiver design and multihop D2D for UAV IoT coverage in disasters. *IEEE Internet of Things Journal* 6 (2): 1803–1815.

7 Ahmed, S., Rashid, M., Alam, F., and Fakhruddin, B. (2019). A disaster response framework based on IoT and D2D communication under 5G network technology. *2019 29th International Telecommunication Networks and Applications Conference (ITNAC)*, pp. 1–6. IEEE.

8 Safdar, G.A., Ur-Rehman, M., Muhammed, M. et al. (2016). Interference mitigation in D2D communication underlaying LTE-A network. *IEEE Access* 4: 7967–7987.

9 Jameel, F., Hamid, Z., Jabeen, F. et al. (2018). A survey of device-to-device communications: research issues and challenges. *IEEE Communication Surveys and Tutorials* 20 (3): 2133–2168.

10 Ali, K., Nguyen, H.X., Vien, Q.-T. et al. (2021). Review and implementation of resilient public safety networks: 5G, IoT and emerging technologies. *IEEE Network Magazine*. 35 (2): 18–25.

11 Feng, D., Lu, L., Yuan-Wu, Y. et al. (2013). Device-to-device communications underlaying cellular networks. *IEEE Transactions on Communications* 61 (8): 3541–3551.

12 Pekka, J., Chia-Hao, Y.U., Cassio, R. et al. (2009). Device-to-device communication underlaying cellular communications systems. *International Journal of Communications, Network and System Sciences.* 3: 169–178.

13 Doppler, K., Yu, C.-H., Ribeiro, C.B., and Jänis, P. (2010). Mode selection for device-to-device communication underlaying an LTE-advanced network. *2010 IEEE Wireless Communication and Networking Conference*, pp. 1–6. IEEE.

14 Ali, K., Nguyen, H.X., Shah, P. et al. (2016). Architecture for public safety network using D2D communication. *2016 IEEE Wireless Communications and Networking Conference Workshops (WCNCW)*, pp. 206–211. IEEE.

15 Moghaddam, J.Z., Usman, M., and Granelli, F. (2018). A device-to-device communication-based disaster response network. *IEEE Transactions on Cognitive Communications and Networking* 4 (2): 288–298.

16 Zhao, N., Lu, W., Sheng, M. et al. (2019). UAV-assisted emergency networks in disasters. *IEEE Wireless Communications* 26 (1): 45–51.

17 Al-Hourani, A., Kandeepan, S., and Jamalipour, A. (2015). Stochastic geometry study on device-to-device communication as a disaster relief solution. *IEEE Transactions on Vehicular Technology* 65 (5): 3005–3017.

18 Nishiyama, H., Ito, M., and Kato, N. (2014). Relay-by-smartphone: realizing multihop device-to-device communications. *IEEE Communications Magazine* 52 (4): 56–65.

19 Ali, K., Nguyen, H.X., Shah, P. et al. (2017). D2D multi-hop relaying services towards disaster communication system. *2017 24th international conference on telecommunications (ICT)*, pp. 1–5. IEEE.

20 Thomas, A. and Raja, G. (2019). FINDER: A D2D based critical communications framework for disaster management in 5G. *Peer-to-Peer Networking and Applications* 12 (4): 912–923.

21 Ali, K., Nguyen, H.X., Vien, Q.-T. et al. (2018). Disaster management using D2D communication with power transfer and clustering techniques. *IEEE Access* 6: 14643–14654.

22 Technical Specification Group Radio Access Network (TSG RAN) (2019). V2X services based on NR; user equipment (UE) radio transmission and reception. http://ftp.3gpp.org//Specs/archive/38_series/38.886/38886-040.zip (accessed 09 October 2021).

23 Wu, Z., Park, V.D., and Li, J. (2015). Enabling device to device broadcast for LTE cellular networks. *IEEE Journal on Selected Areas in Communications* 34 (1): 58–70.

24 Maddio, S., Cidronali, A., Passafiume, M. et al. (2015). Interference cancellation for the coexistence of 5.8 GHz DSRC and 5.9 GHz ETSI ITS. *2015 IEEE MTT-S International Conference on Microwaves for Intelligent Mobility (ICMIM)*, pp. 1–4. IEEE.

25 Khoueiry, B.W. and Soleymani, M.R. (2015). A novel coding strategy for device-to-device communications. *2015 12th Annual IEEE Consumer Communications and Networking Conference (CCNC)*, pp. 200–205. IEEE.

26 Safdar, G.A., Ur-Rehman, M., Muhammad, M. et al. (2016). Interference mitigation in D2D communication underlaying LTE-A network. *IEEE Access* 4: 7967–7987.

27 Gu, J., Bae, S.J., Choi, B.-G., and Chung, M.Y. (2011). Dynamic power control mechanism for interference coordination of device-to-device communication in cellular networks. 2011 3rd International Conference on Ubiquitous and Future Networks (ICUFN), pp. 71–75. IEEE.

28 e Silva, C.F.M., Batista, R.L., da Silva, J.M.B. et al. (2015). Interference mitigation using band selection for network-assisted D2D communications. *2015 IEEE 81st Vehicular Technology Conference (VTC Spring)*, pp. 1–5. IEEE.

29 Bao, P., Yu, G., and Yin, R. (2013). Novel frequency reusing scheme for interference mitigation in D2D uplink underlaying networks. *2013 9th International Wireless Communications and Mobile Computing Conference (IWCMC)*, pp. 491–496. IEEE.

30 Qin, Y., Ding, M., Zhang, M. et al. (2014). Relaying robust beamforming for device-to-device communication with channel uncertainty. *IEEE Communications Letters* 18 (10): 1859–1862.

31 Ge, X., Zi, R., Wang, H. et al. (2016). Multi-user massive MIMO communication systems based on irregular antenna arrays. *IEEE Transactions on Wireless Communications* 15 (8): 5287–5301.

32 Zugno, T., Drago, M., Giordani, M. et al. (2020). NR V2X communications at millimeter waves: an end-to-end performance evaluation. *arXiv preprint arXiv:2005.10148*.

9

Disaster Management Using D2D Communication With Power Transfer and Clustering Techniques

Kamran Ali[1], Huan Nguyen[1], Aboubaker Lasebae[1], Anum Tanveer[1], Purav Shah[1], Mohsin Raza[2], Bushrah Naeem[3], and Tahera Kalsoom[4]

[1]*Computer Science Department, Middlesex University, London, United Kingdom*
[2]*Department of Computer Science, Edge Hill University, Ormskirk, United Kingdom*
[3]*Department of Computer Science, Balochistan University of Information Technology, Engineering, and Management Sciences (BUITEMS), Quetta, Pakistan*
[4]*School of Computing, Engineering and Physical Sciences, University of West of Scotland, Paisley, United Kingdom*

Communication is one of the biggest problems in disaster situations. Many people around the world are adversely affected by various unforeseen disasters such as earthquakes, tsunami, floods, and torrential rains. Although disasters differ according to the nature and the context in which they occur, their management faces almost the same and repetitive problems which are mainly related to the complexity of the field, the interoperability issues, and the sociocultural aspects. Effective emergency and natural disaster communication system depends on the efficient mission-critical voice and data communication between first responders and victims [1]. The limitations which restrict the realization of information communications are [2]: (i) communication infrastructures are not available; (ii) loss of energy/power; (iii) low reliability and availability of network, and (iv) limited resources and limited services.

A disaster communication architecture based on device-to-device (D2D) communication is discussed in this chapter. D2D can achieve higher system throughput by taking advantage for short distance, direct communications, and multi-users diversity. Furthermore, integrating D2D with the cellular network offers the prospect of a spectrum-efficient, energy-efficient, and secure solution. The proximity-based services of D2D technology also satisfy the need of public safety networks (PSNs). In particular, we consider a D2D-based energy-harvesting cellular network where the relay and user equipment relay (UER) harvest radio frequency (RF) energy from a base station (BS) and use the harvested energy for the D2D communications. Further, we employ user-cooperation-based clustering scheme to minimize the average power consumption of user equipments (UEs).

Interference Mitigation in Device-to-Device Communications, First Edition.
Edited by Masood Ur Rehman, Ghazanfar Ali Safdar, and Mohammad Asad Rehman Chaudhry.

9.1 D2D Communication in Disaster Management

Situations like disasters (earthquakes/tsunamis) need power saving, less energy-consuming devices, and networks. Certainly, energy consumption is a key area nowadays in PSNs. In addition to this, the breakdown of essential communications is one of the most commonly shared characteristics of all disasters. Whether partial or complete, the failure of telecommunications infrastructure leads to unavoidable loss of life by causing delays and faults in emergency response of disaster relief. In many situations, the reason for the disconnection of communication networks is due to the breakdown of power networks. Although some small communication devices run on batteries, their power may run out quickly. Therefore, recovery of power supply is critical for disaster recovery.

In order to cope with this critical issue, energy harvesting (EH) is capable of powering communication devices and networks with the energy harvested from the environment. Recently, the EH has become an appealing solution to prolong the lifetime of wireless networks. The EH potentially has an unlimited energy supply from the environment in the form of renewable energy sources like solar and wind energy. Therefore, the research in renewable energy has drawn a great attention particularly in cellular communication [3]. In addition, in disaster situations, energy efficiency is even more important in terms of robustness and availability than in normal conditions. Solar panel and wind turbine do not justify the scalability and reliability to situation. Therefore, ambient energy radio signals can be viable new resource for wireless energy harvesting (WEH) where the RF systems have the capability of converting the received signals into energy source [4]. The system becomes a promising solution to energy-constrained wireless networks. As the energy is particularly limited in emergency situations, RF energy can become an excellent choice for the power and information exchange.

Simultaneous wireless information and power transfer (SWIPT) has developed as a promising technology, allowing the transmission of data and energy together to a receiver equipped with an RF energy-harvesting circuitry. Moreover, SWIPT is capable of delivering suitable and continuous energy provisions to the wireless networks. This accordingly motivates us to investigate the SWIPT and its connotation in disaster communication systems because of its significant attentions and also to be considered as an emerging technology. In a communication system with SWIPT capabilities, information and power transfer are carried out over wireless medium simultaneously. Two practical schemes for SWIPT, namely *time switching* (TS) and *power splitting* (PS), were proposed in [5].

This chapter details a framework that apply time switching protocol at relay to provide energy and information to further strengthen the life of relay-based network during disaster events based on our work presented in [6]. Mode selection strategy is adopted to analyze the suitability of D2D communication and the employment of the UER during disaster to harvest energy from the relay via BS so that the energy is sustained for D2D communication. This research contributes to the understanding of the effects of the network parameters on the outage probability of the D2D mode. A novel cooperative D2D energy-harvesting clustering network for disaster management is also developed to enhance performance, reliability, and energy efficiency. The network is shown to be resilient and scalable for large number of users. It helps minimize impact of the limitation of resources and services to overcome major problems during disaster events such as power loss, traffic congestion, and network capacity.

9.2 D2D Communication in Disaster Management: Key Considerations

Recently, it has been observed in Italy, Nepal, and New Zealand that due to the earthquakes most of the existing networks were completely destroyed in the disaster areas. This affected local people significantly made the first responders' tasks very difficult. A large number of victims were trapped in the disaster zones for several days. Such big-scale disasters require a PSN system that can operate efficiently in disaster situations. The research community has attempted to propose various disaster-resilient networks on the basis of earlier disasters experiences and related projects around the world. The common goal of all the efforts is to design network architectures and provide solutions that are resilient to disasters.

During those above disaster situations, Facebook activated its Safety Check feature to enable people to give quick safety status updates to their family and friend during the disaster [7]. Popular social media tools like Twitter and Facebook were flooded with messages for requests and help. It was observed that on the disaster sites, local people had little capacity to gather and arrange the large amount of information coming in. This has led to the demand to investigate the limitations of the cellular networks to handle the traffic in those critical situations. Fortunately, D2D communication effectively uses the radio resources with the goal of serving a large number of affected people to collect information from different nodes in the disaster zone [8]. In [9], the authors developed a prototype, namely Relay by Smartphone, of D2D relaying smartphone that enables sending out emergency messages from disconnected areas and sharing of information between people gathered in evacuation centers. In [10], a novel D2D-based messaging solution to overcome the UE power limitation issues faced by cellular radio-access technologies in disaster situations was presented. The proposed D2D messaging mechanism was compared with the default random-access channel (RACH)-based messaging mechanism. However, in this work, the RACH-based power consumption was not explicitly modeled for various radio-access technologies (e.g. Global System for Mobile Communications [GSMs], Universal Mobile Telecommunications Systems [UMTSs], and Long-Term Evolution [LTE]), and therefore further study is needed to evaluate the power consumption.

According to the requirement of PSN, wireless networks are ideal choice for disaster relief operations as wireless networks do not need any preexisting infrastructure to be set up and are easy to operate in critical situations. However, energy management is a big concern for such infrastructure. Therefore, some recent studies have considered EH through the RF signals in wireless cooperative networks. In [11], the authors considered a multiple-input multiple-output (MIMO) relay system and studied the trade-off between the energy transfer and the information rate to achieve the optimal source and relay precoding. However, they assumed that the relay has its own internal energy source and does not need external charging. In contrast to [11], in our work, we consider that the relay also relies on external wireless charging through the RF signal from the source node. The authors of [12] presented hybrid power transfer architecture with power relay. Airships, helicopters, and balloons are used to transfer power wirelessly through inductive power transfer to big communication devices and through microwave power transfer to small communication devices. Differently, we use efficient and practically workable BS and relay to transfer energy wirelessly to the destinations.

The work in [13] investigated renewable energy-enabled base station (REBS) and pre-equipped energy-harvesting devices, performing in wireless mesh network fashion, for post-disaster communication scenarios. Particularly, the authors focused on optimizing data traffic throughput with lowest weighted energy consumption, in which they proposed an off-line energy-efficient scheme using the expectation of traffic demands. In addition, for better energy consumption and management in future cellular networks, the authors in [14] proposed a new algorithm to help in reducing energy consumption of LTE heterogeneous network (HetNet) using macro- and micro-base stations and cell zooming. The solution is carried out through introducing new cell sizes (micro, pico, and femto cells) with various capacities, coverage areas, and lower power consumption by operating all together existing base stations in a HetNet. However, the existing scholarly works have focused mainly on reducing the energy consumption of BS via HetNet and cell zooming, whereas in our approach, we develop extended communication links in the disaster area by adopting EH-D2D technique and clustering method. In [15], resource unit concept called Movable and Deployable Resource Unit (MDRU) was presented by Nippon Telegraph and Telephone (NTT) Corp. The idea of MDRU is to transport complete resource unit to the disaster site and deploy recovery network. However, the cost of making an MDRU is very high, and the MDRU deployment may not be practical considering that spectrum and energy resource in a disaster area are significantly limited while the demand of network connectivity, capacity, and power often increases with time during and after disaster.

In [16], the authors proposed a stand-alone eNode-B architecture, which deploys its own integrated virtual evolved packet core (EPC) to ensure service without backhaul connection. The proposed stand-alone eNode-Bs are also designed to establish backhaul connection with each other to extend the coverage without need of a central EPC structure. The concept of eNode-B not only offers better interoperability but also increases functionality in terms of transmitting data, especially in emergency situations and disaster scenarios. However, the work in [16] did not address the power consumption issue during the disaster events. In [17], the effect of the

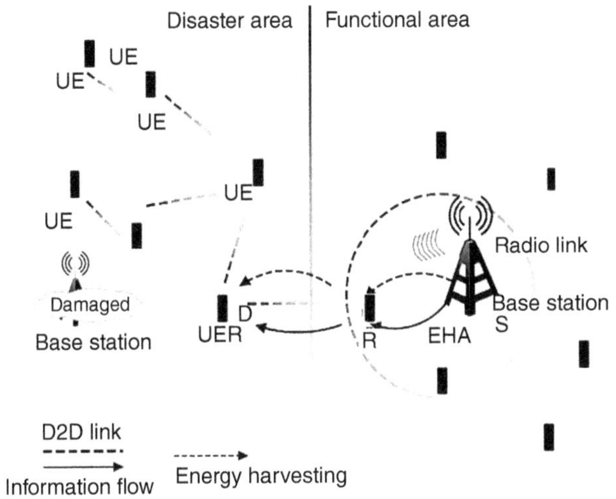

Figure 9.1 Energy harvesting concept in disaster situation [18].

relay mobility was addressed, where the authors considered the coverage and capacity extensions by mobile relays, and the influence of mobility on the probability of route establishment and the expected availability duration. However, this work limits their scope to point-to-point communication with single cell in idealized circular coverage area, whereas here in our work we address the communication links across the entire network of multi-cell, multi-hop where the network coverage is extended beyond a single cell using D2D relay links.

9.3 D2D Disaster Management System Architecture

Mobile communications with D2D proximity have unique topographies which are valuable in disaster and emergency situations. As long as BS is up and running or an ad hoc process can manage to provide reliable communications (relay system), wireless links can be easily established between users within its coverage area or with extended options.

We consider a public safety scenario as shown in Figure 9.1, where a source (i.e. BS) with a fixed energy supply desires to transmit its information to a destination located in the out of coverage area. Due to the barrier between the source and destination or direct link distance, the source cannot directly transmit its signals to the destination; thus, it asks a relay to assist its information transmission via D2D communication proximity services, since destination is in a disaster area where direct communication is not possible. However, because of the selfish nature, or lack of energy supply, the relay needs to harvest energy before being able to help.

Throughout this work, the following set of assumptions are considered.

Assumption 9.1 A wireless D2D communication system is considered, where the information is transferred from source S to destination node D, through intermediate relay node R as shown in Figure 9.1. The destination node D can then act as a *(UER)* node in another hop in the disaster area. Further, UER acts as a cluster head *(UERCH)* in clustering formation phase.

Assumption 9.2 The intermediate node R is an energy-constrained node. It first harvests energy from the S and then uses the harvested energy as a source of transmit power to forward the source information to the destination.

Assumption 9.3 Relay node is selected to forward the signals and energy to the destination only if it has sufficient energy or is able to harvest energy from the source. Among the different relaying protocols, Decode and Forward (DF) scheme is chosen at the relay node.

9.3.1 Time Switching-Based Protocol

The TS-based protocol is adopted at the relay, where T denote the block time in which a certain block of information is transmitted from the *source* node to the *destination* node with TS ratios α_1, α_2, and α_3 where $\alpha_1 + \alpha_2 + \alpha_3 = 1$. During the first-time slot with duration of $\alpha_1 T$, the source transfers energy to the relay. In the second-time slot $\alpha_2 T$, the source transmits signal to the relay while in $\alpha_3 T$ the relay forwards signals to the destination. There is no direct link between the source and the destination node.

Assume that in relay network, the total bandwidth is divided into N orthogonal subcarriers, $n \in N = 1,2,3,\dots,N$, and the network comprises two wireless links i.e. source to relay and relay to destination links as h_n^{SR} and h_n^{RD}, respectively. We further assume that $h_n^{\mathrm{SR}} \propto d_{\mathrm{SR}}^{\alpha}$ and $h_n^{\mathrm{RD}} \propto d_{\mathrm{RD}}^{\alpha}$ where α is the path-loss exponent, and d_{SR} and d_{RD} are the distances between source and relay and between relay and destination, respectively. The energy harvested at the relay during the first-time slot can be expressed as [5]:

$$E = \alpha_1 T\zeta \sum_{n=1}^{N} p_n^{S,1} |h_n^{\mathrm{SR}}|^2 \tag{9.1}$$

where $p_n^{S,1}$ represents the transmit power of the source (i.e. BS) over nth subcarrier for energy transfer and ζ denotes EH efficiency. To maximize harvested energy at the relay, the source should assign all available power over the subcarrier that has a maximum channel gain. Therefore, we have

$$E = \alpha_1 G \tag{9.2}$$

where

$$G = T\zeta P \max_n |h_n^{\mathrm{SR}}|^2 \tag{9.3}$$

and P represents the maximum transmit power of the source for both energy and information transmission. We have $P \geq \sum_{n=1}^{N} p_n^{S,1}$. Then, the source transmits signals to the relay over N subcarrier in the second-time slot. After receiving the signals, relay decodes the signals, redistributes the decoded signals over different subcarriers, and forwards the signals to destination. Therefore, the maximum achievable end-to-end data rate of DF relay network is obtained as [19]:

$$C = \min\left[\alpha_2 \sum_{n=1}^{N} \log_2(1 + p_n^{S,2}\gamma_n^{\text{SR}}), \alpha_3 \sum_{n=1}^{N} \log_2(1 + p_n^R\gamma_n^{\text{RD}})\right] \tag{9.4}$$

where $p_n^{S,2}$ and p_n^R represent the transmit power of source and relay over nth subcarrier for information transmission, respectively; $\gamma_n^{\text{SR}} = |h_n^{\text{SR}}|^2/\sigma_R^2$ and $\gamma_n^{\text{RD}} = |h_n^{\text{RD}}|^2/\sigma_D^2$. Here, noise powers over each subcarrier at relay and destination are represented by σ_R^2 and σ_D^2, respectively. It is also noticed from [20] and [21] that the harvested energy in the first-time slot should be larger than or equal to the consumed energy for information transmission at the relay, i.e.

$$E \geq \alpha_3 T \sum_{n=1}^{N} p_n^R \tag{9.5}$$

Note that there are possibly numerous nodes between source and destination, which are candidates to perform as a relaying node. Here we need to define the best relay node before information and energy transfer.

Relay node selection has already been discussed in our earlier work. It is assumed that the BS knows every channel in the field. So, it can simply determine which node can act better in the area. In order to harvest energy, the received power of a relay from its nearby BS should not be small. We define the area around the BS called *energy-harvesting area (EHA)* which is circle with the radius R_{ha} centered at BS as shown in Figure 9.1 and the R_{ha} is given by:

$$R_{\text{ha}} = \left(\frac{\zeta P_{\text{bs}}}{B_w}\right)^{1/v} \tag{9.6}$$

where $\zeta \in (0,1)$ is the energy-harvesting efficiency factor of UE, P_{bs} transmit power of the BS, B_w is the energy-harvesting threshold to activate the EH circuit, and v is path-loss exponent in EH transfer link.

9.3.2 Network Configuration

As shown in Figure 9.1, we consider a cellular network which consists of BS capable of performing wireless power transfer and relay that can harvest energy from their nearby BS. We assume that the distribution of the BS in the network follows homogeneous Poisson point process (PPP) Φ_{bs} with spatial density λ_{bs} and UEs are also distributed according to a homogeneous PPP Φ_{ue} with spatial density λ_{ue}, while relay spatial density is λ_r. The transmit powers of a BS, relay, and UE are P_{bs}, P_r, and P_{ue}, respectively. When a relay has sufficient harvested energy, then it can help the BS to forward the information to another hop in the nonfunctional area node in DF manner and with signal-to-interference ratio (SIR). The node acts as a UER, and the communication between a relay and the destination UE is taken place between two user devices, which is D2D communication. As we know that UEs cannot communicate with a BS for their own data, we also assume that each UER receives the data for itself with probability p_{rc} and p_r is the probability that an UE can support other UEs as an UER. Moreover, when an UER does not need to receive its own data and has the amount of harvested energy exceeding $E_{\rm ue} = NB_w$, the UER can help another UE using $E_{\rm ue}$. Load $\rho_{\rm bs}$, as BS using the same spectrum at any random time slot, is expressed as [22]:

$$\rho_{\rm bs} = \frac{p_{\rm rc}\lambda_r}{\lambda_{\rm bs}N} \tag{9.7}$$

9.3.3 Outage Probability for Mode Selection

Applying relaying techniques to D2D communication scenarios has attracted much attention recently due to its ability to enhance D2D coverage and reliability. End-to-end outage probability is investigated here to confirm whether D2D technology could be a preferable option or not in our scenario (mode selection). First, we determine the outage probability of an UER that operates in D2D mode. The distance between BS and R is d_1 and the distance between R and an intended UER is d_2, and the outage probability of D2D mode can be presented as:

$$P_{\rm out} = 1 - \exp\left\{-\xi(\theta_d, \alpha)\left(\rho_{\rm bs}\lambda_{\rm bs}d_1^2 + \frac{p_r\lambda_r}{N}d_2^2\right)\right\} \tag{9.8}$$

where α is path-loss exponent in data link and θ_d is the SIR threshold for D2D mode transmission and $\xi(\theta_d, \alpha)$ is given as,

$$\xi(\theta, \alpha) = \frac{2\pi^2}{\alpha}\csc\left(\frac{2\pi}{\alpha}\right)\theta^2/\alpha \tag{9.9}$$

In the D2D mode transmission, the outage occurs when at least one of the two links (BS to R, and R to UER) does not achieve the target SIR θ_d. Consider that in Figure 9.1 the BS locates at (x_s, y_s), R locates at (x_r, y_r), and UER locates at (x_d, y_d), then we can have $d_1^2 = (x_r - x_s)^2 + (y_r - y_s)^2$ and $d_2^2 = (x_d - x_r)^2 + (y_d - y_r)^2$. The outage probability in (9.8) can now be rewritten as:

$$P_{\rm out} = 1 - \exp\left\{-\rho_{\rm bs}\lambda_{\rm bs}\xi(\theta_d, \alpha)f(x_r, y_r)\right\} \tag{9.10}$$

where

$$\begin{aligned} f(x_r, y_r) &= |(x_s - x_r)|^2 + |(y_s - y_r)|^2 \\ &\quad + \Lambda||(x_r - x_d||^2 + \Lambda|(y_r - y_d)|^2) \end{aligned} \tag{9.11}$$

and Λ is given as:

$$\Lambda = \frac{p_r\lambda_r}{N\rho_{\rm bs}\lambda_{\rm bs}} \tag{9.12}$$

The optimal location of R that minimizes $P_{\rm out}$ is obtained by:

$$\begin{aligned} (x_r^o, y_r^o) &= \arg\min_{\{x_r, y_r\}} P_{out} \\ &= \arg\min_{\{x_r, y_r\}} f(x_r, y_r) \end{aligned} \tag{9.13}$$

By taking partial differentiation of $f(x_r, y_r)$ with respect to x_r and y_r separately and equate it to zero, we can achieve the optimal location of R as follows:

$$(x_r^o, y_r^o) = \left(\frac{x_s + \Lambda x_d}{1 + \Lambda}, \frac{y_s + \Lambda y_d}{1 + \Lambda}\right) \tag{9.14}$$

Now, using the optimal location of R, the outage probability can be represented as:

$$P_{\text{out}}(x, d_o) = 1 - \exp\left\{-\rho_{\text{bs}}\lambda_{\text{bs}}\xi(\theta_d, \alpha)\left(\frac{\Lambda x^2}{1+\Lambda} + (1+\Lambda)d_o^2\right)\right\} \tag{9.15}$$

where x is the distance between a source and the UER, and the d_o is the distance between the optimal R location (x_o, y_o) and the actual R location. From (9.15) and (9.17), we can see that R is a circle centered at (x_o, y_o). The radius of this circle can be used to determine whether the outage probability is satisfactory. If it is then, we can conclude that the UER is in the D2D selection mode and will use the nearest R from the optimal R location for its further communication.

This transmission mode selection can be useful to exploit the D2D energy-harvesting network (EHN) communication efficiency. Therefore, it is useful to study the outage probability of the D2D-EHN by considering the transmission mode selection and the EH as follows:

$$\begin{aligned}P_{\text{out}} = {} & \frac{\rho_{\text{bs}}\lambda_{\text{bs}}\xi(\theta_d, \alpha)(1+\Lambda)}{\lambda_r\pi + \rho_{\text{bs}}\lambda_{\text{bs}}\xi(\theta_d, \alpha)(1+\Lambda)} \\ & \times \frac{\lambda_r\kappa^2\pi + \rho_{\text{bs}}\lambda_{\text{bs}}\varphi(\theta_b, \alpha)}{\lambda_{\text{bs}}\pi + \lambda_r\kappa^2\pi + \rho_b\lambda_{\text{bs}}\varphi(\theta_b, \alpha)} \\ & + \frac{\lambda_r\pi}{\lambda_r\pi + \rho_{\text{bs}}\lambda_{\text{bs}}\xi(\theta_d, \alpha)(1+\Lambda)} \\ & \times \frac{\xi(\theta_d, \alpha)\rho_{\text{bs}}\lambda_{\text{bs}}\Lambda}{\lambda_{\text{bs}}\pi(1+\Lambda) + \xi(\theta_d, \alpha)\rho_{\text{bs}}\lambda_{\text{bs}}\Lambda}\end{aligned} \tag{9.16}$$

where

$$\kappa = \sqrt{\left(\frac{\varphi(\theta_b, \alpha)}{\xi(\theta_d, \alpha)} - \frac{\Lambda}{1+\Lambda}\right)\frac{\Lambda}{1+\Lambda}} \tag{9.17}$$

and

$$\begin{aligned}\lambda_r &= \lambda_{\text{ue}} p_a (1 - p_{\text{rc}}) \\ &= \frac{\lambda_{\text{ue}} p_r^{-1}(1 - p_{\text{rc}})\left(1 - e^{-\pi\lambda_{\text{bs}}R_h^2}\right)}{1 - e^{-\pi\lambda_{\text{bs}}R_h^2} + N}\end{aligned} \tag{9.18}$$

and

$$\begin{aligned}\varphi(\theta, \alpha) &= \frac{2\pi\theta^{2/\alpha}}{\alpha}\mathbb{E}\left[h^{2/\alpha}\left(\Gamma\left(-\frac{2}{\alpha}, \theta h\right) - \Gamma\left(-\frac{2}{\alpha}\right)\right)\right] - \pi \\ &= \frac{2\pi\,{}_2F_1\left(1,1+\frac{2}{\alpha}, 2+\frac{2}{\alpha}, -\frac{1}{\theta}\right)}{(2+\alpha)^{\theta}} - \frac{2\pi^2\theta^{2/\alpha}}{\alpha}\csc\left(\frac{2\pi}{\alpha}\right) - \pi.\end{aligned}$$

Here, h is the channel gain which experiences Rayleigh fading, and ${}_2F_1(a, b; c, x)$ is the hypergeometric function. We assess the performance of the system and prove that D2D is the favorable option for communication in disaster situations.

As shown in Figure 9.2, we observe that an optimal efficiency value exists and shifts toward zero as P_{bs} becomes larger. For D2D EHN with $P_{\text{bs}} = 3$ W, P_{out} reduces exponentially as the EH efficiency increases, but with $P_{\text{bs}} = 5$ W, P_{out} reduces exponentially till that EH efficiency is equal to 0.5. After that, P_{out} increases steadily.

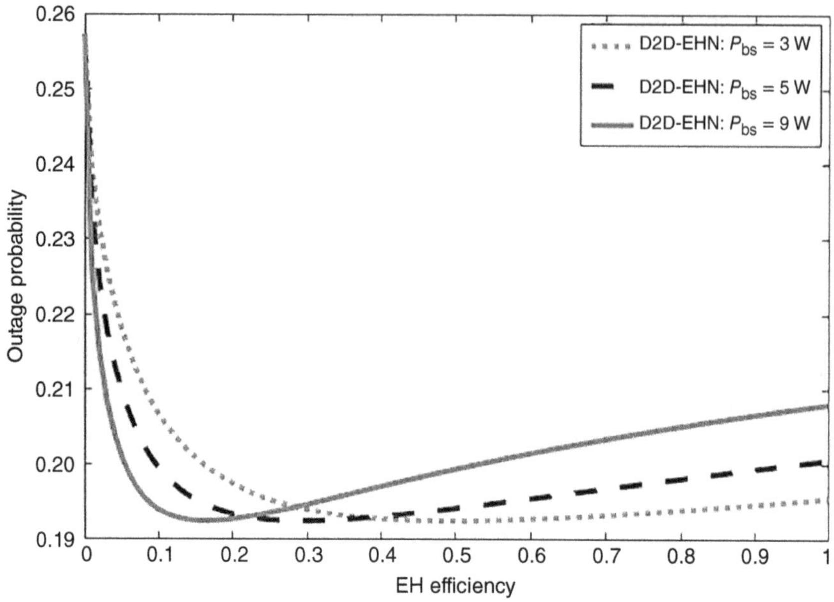

Figure 9.2 Outage probability as a function of energy efficiency for different values of BS power (P_{bs}) [18].

Moreover, with P_{bs} = 9 W, P_{out} decreases dramatically till the EH efficiency value of 0.2 and then increases linearly subsequently. As a result, there are more chances to adopt UEs to select D2D mode for the purpose to reduce the outage probability. For larger P_{bs}, we observe that the UER density increases dramatically with ζ, which means that it leads to increasing the number of UERs, and each UE has a high chance to communicate in D2D mode.

Figure 9.3 shows the outage probability (P_{out}) as a function of p_{rc} (i.e. probability by which an UER receives data for itself). When a UER is receiving its own data from the R, it is incapable of serving as an UER. We observe that P_{out} is increasing when p_{rc} increases. Moreover, using D2D communications reduces P_{out}, especially with high EH efficiency. Higher values of EH efficiency factor reduce p_{rc} which increases the effect on P_{out}, and it stabilizes around the value of 0.07. As shown in Figure 9.4, outage probability is a function of P_r (i.e. probability that a UE can support other UEs as a UER) for different values of EH efficiency. Outage probability increases as P_r increases but in different manners, exponentially in the absence of D2D communications and linearly when D2D communications applies. This makes the prediction of the D2D outage probability possible. When a UER has harvested sufficient energy and served as a UER, it can then be solicited by other UEs to relay their data and connection.

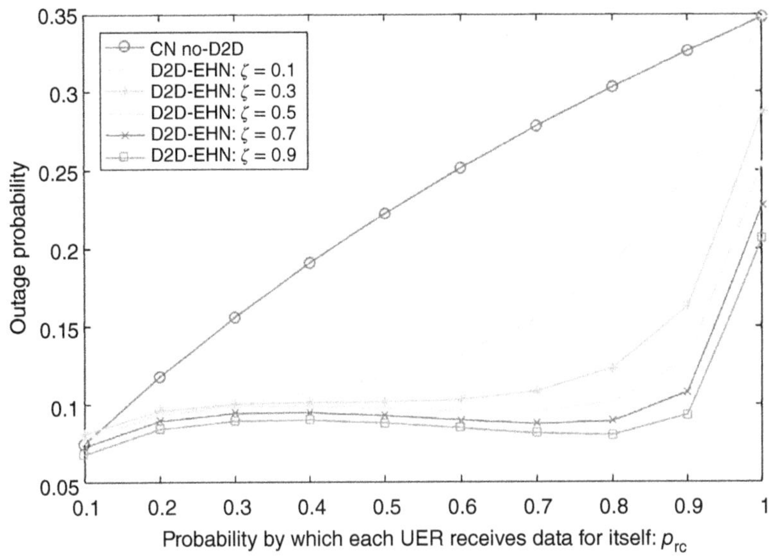

Figure 9.3 Outage probability as a function of p_{rc} for different values of ζ [18].

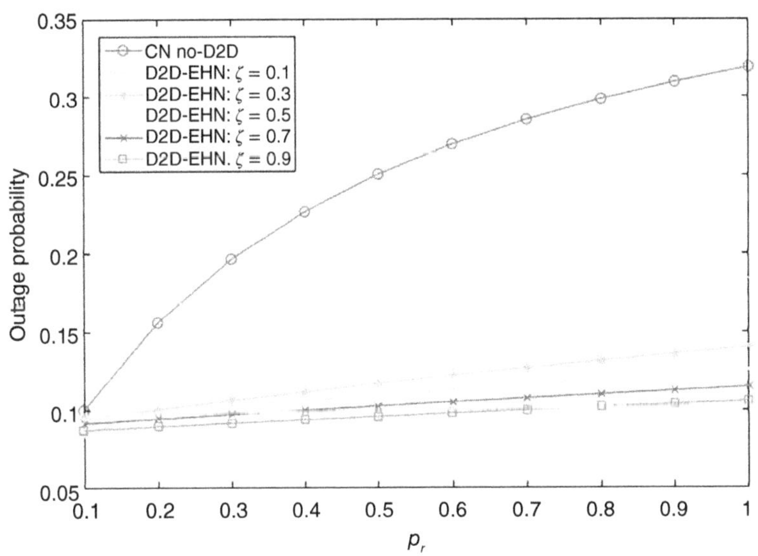

Figure 9.4 Outage probability as a function of p_r for different values of ζ [18].

9.4 Power Transfer Using Relaying and Clustering in D2D Disaster Management

9.4.1 System Model

In Section 9.3, we aimed to send signals from functional area to nonfunctional area with the help of a network relay. We also implemented energy-harvesting technique so that the relay could be active in the field and further passes the energy and information into the disaster area as per requirement of the system. Now, in this section, we are presenting a novel approach by using clustering techniques in D2D relay mode to facilitate communication of multiple users that are affected within the disaster area. From the technical perspective, exploiting the nature of proximity may provide multiple benefits in disaster situations like:

- D2D may utilize high data rate and low end-to-end delays.
- Compared to normal cellular communication, direct communication saves energy and improves utilization ratio.

Since network-assisted D2D communications take advantage of the cellular infrastructure presences, we propose to build on the D2D underlay concept but extend it in such a way that allows infrastructure/infrastructure-less operation. According to our scenario, high-end capability UE can take over some of the radio-access network (RAN) functionalities when one or more BS becomes dysfunctional. Such functionalities include providing synchronization signals and acting as a UER or cluster head (CH). We will use in this work the concept and functionality of user equipment relay and cluster head (UERCH). We have observed that the capability of nodes to become a UERCH must be taken into account (e.g. available transmit power, support spectrum or synchronization or radio resource management capability). However, network coverage distinguishes two types of UEs:

Category 1 (Cat-1): The UEs are capable of becoming UERCH, directing the D2D links and managing the resource usage among a group of D2D devices (UEs) associated with them.

Category 2 (Cat-2): The UEs are devices that can only act as cluster member and according to disaster situation they are out of network coverage area and controlled by appropriate Category-1 UEs.

UERCH selection has already been discussed in our previous work [23] where complete flowchart is presented to differentiate between nodes of Cat-1 and Cat-2 category. Furthermore, UERCH depends on the outage probability derived in Section 9.3.3. UERCH node proposed in our architecture depends on nodes broadcasting so-called beacon signal on peer discovery resource (PDR). The authors of [24] evaluated the impact of network assistance on the performance of D2D discovery algorithms in terms of discovery probability, discover time, and consumed energy. In [25] grouping strategies, selection of PDRs, beacon signaling, and the setting of beacon transmission probability in each time slot with beacon transmission power were discussed. The authors also highlighted the energy, spectral efficiency, and the capability of dynamically reconfiguring the network due to mobility, changing radio conditions, and nodes joining/leaving the network.

The key of our approach is to elect which user device should act as UERCH as we are establishing communication link via network relay, so the following points must be considered in reality to select the UERCH:

- *Capacity*: UERCH must have certain functions like dual-mode function, which means that they can work in both low- and high-power mode.
- *Network coverage*: If the device is able to achieve network coverage, other devices are capable to connect to the network through it. Besides, the network can assist D2D communication and make the system more efficient.
- *Mobility*: If user moves fast, it can easily move out of the current cluster, and the stable situation is changed and a UERCH reselection is needed which costs computation and time. Thus, slowly moving or static devices are more suitable to be a UERCH.

9.4.2 Performance Evaluation

In order to cope with disaster situations, a novel cooperative disaster D2D clustering is implemented. For better performance, power transfer relay and D2D clustering methods are adopted to provide novel, robust, and stable solution at the time of critical situation. We consider a scenario, in which we successfully establish link from functional area's BS to nonfunctional area's devices with the help of a network relay. A number of devices can then form a cluster (i.e. coalition) with one device acting as a UERCH and the rest of the devices in the cluster are called cluster members (CMs). Moreover, within each cluster, D2D communications are adopted to perform content distribution. Data content is sent from BS via R using long-range link, as shown in Figure 9.5, to the UERCH, and then each UERCH send the content to its particular CM via short-range link.

All the devices operate in D2D mode. At the time of communication, each CM receives the content from a single-source UERCH. We assume that all nodes can form coalition (see Algorithm 1) where the energy consumption in the coalition is lower than the sum of the individual energy consumptions of the coalition members [26, 27]. Meanwhile, the UERCH also communicates with the relay on behalf of all CMs. Assume that there are K clusters to be formed, $C_1, C_2, \dots, C_K$, as shown in Figure 9.5. Within each cluster C_k $(k = 1, 2, \dots, K)$, there are $I_k + 1$ nodes: one UERCH node $n_{k,0}$, and I_k other CM nodes n_{k,i_k}, $i_k = 1, 2, \dots, I_k$.

9.4.2.1 Energy Calculation

Consider a communication link between two nodes $n_{k,u}$ and $n_{k,v}$ $(u, v = 1, 2, \dots, i_k$ and $u \neq v)$ with cluster C_k. The time needed to transmit a data content of size S_T bits

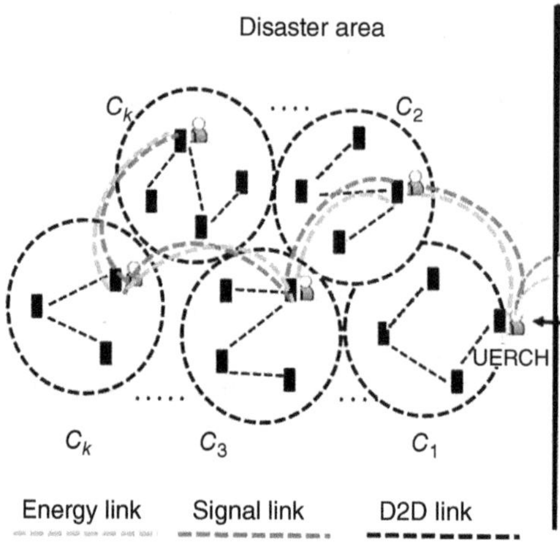

Figure 9.5 Combined system model framework for disaster recovery communication using D2D, clustering and EH [18].

on this link having an achievable rate R_{uv} bps is given by S_T/R_{uv}. Denoting the power drained from the battery of node $n_{k,v}$ to receive the data from node $n_{k,u}$ by $P_{\text{Rx},uv}$, then the energy consumed by $n_{k,v}$ to receive the data from $n_{k,u}$ is given by $S_T P_{\text{Rx},uv}/R_{uv}$. Likewise, representing by $P_{\text{Tx},uv}$ the power drained from the battery of node $n_{k,u}$ to transmit the data to $n_{k,v}$, and then the energy is consumed by $n_{k,u}$ to transmit the content to $n_{k,v}$ is given by $S_T P_{\text{Tx},uv}/R_{uv}$, where $P_{\text{Tx},uv}$ can be expressed as:

$$P_{\text{Tx},uv} = P_{\text{Tx}_{ref},uv} + P_{t,uv} \tag{9.19}$$

where, $P_{t,uv}$ is transmit power over the air interface on the link between nodes and $P_{\text{Tx}_{\text{ref}},uv}$ is the power consumed by the circuitry of nodes during transmission. Denote E_{C_k} as the total energy consumed by cluster C_k, which can be expressed as:

$$E_{C_k} = S_T \sum_{u \neq v, u, v = 1,2,\ldots,J_k} \left(\frac{\Gamma_k P_{\text{Tx},uv} + P_{\text{Rx},uv}}{R_{uv}} + \frac{P_{\text{Rx},u}}{R_u} \right) \tag{9.20}$$

where the first term links to the energy consumed by node $n_{k,u}$ to receive the data from R on the large link called cellular link; the second term corresponds to energy consumed by the node $n_{k,u}$ to transmit the data to the other nodes in its cluster on the short link via D2D communication and the last term corresponds to the energy consumed by the nodes to receive their data from node $n_{k,u}$ on the SR. The variable Γ_k is used to differentiate between unicasting and multicasting. In fact, in the uplink process, each node has different data to transmit. Hence, only unicasting is used but on the other hand, the same data are transmitted to the member of each cluster in the downlink case.

Considering a single cluster only, the harvested energy (in (9.1)) should not be smaller than the energy consumption (9.20), leading to:

$$E \geq E_{C_k}$$

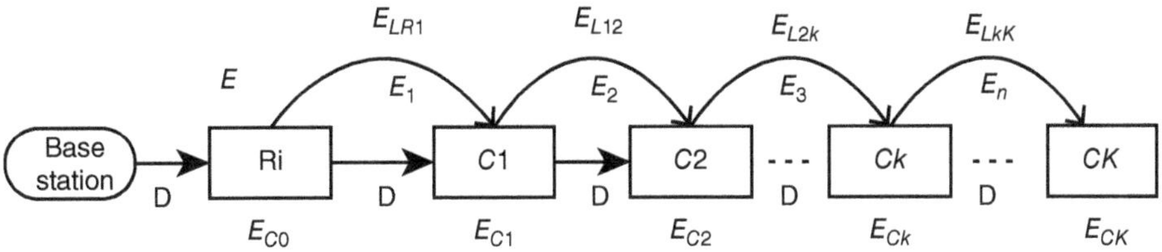

Figure 9.6 Energy loss during transfer of energy [18].

$$\alpha_1 T\rho \sum_{n=1}^{N} p_n^{S,1} |h_n^{\mathrm{SR}}|^2 \geq E_{C_k}$$

$$\sum_{n=1}^{N} p_n^{S,1} |h_n^{\mathrm{SR}}|^2 \geq \frac{E_{C_k}}{\alpha_1 T\rho} \tag{9.21}$$

Assuming power of each subcarrier is the same, (i.e. $p_1^{S,1} = p_2^{S,1} = p_N^{S,1} = p^{S,1}$) we should have

$$p \geq \frac{E_{C_k}}{\alpha_1 T\rho \sum_{n=1}^{N} |h_n^{\mathrm{SR}}|^2} \tag{9.22}$$

Here, $p^{S,1}$ represents the power of single subcarrier required for energy recovery as a function of number of devices. This power refers to the necessary single subcarrier power in such a way that harvested energy from the source in the relay is greater or equal to consumed energy to transmit the signal from relay to destination.

Now, we consider multiple clusters to be formed. Each cluster will transfer energy and communicate with the next cluster in a serial multi-hop manner, as shown in Figure 9.6. In addition to the energy consumption E_{C_k}, there will be energy loss when transferring between one cluster to the next, denoted by $E_{L_{(k-1),k}}$.

We therefore propose a coalition formation algorithm (see Algorithm 1) to form the clusters in the most energy-efficient way. In particular, the proposed algorithm is to form the clusters among all the UEs in a way to reduce the average energy consumption. As soon as a UE decides to enter/form a cluster, it enters a binding agreement with the other users within the coalition and then considers the benefit of the coalition above their individual benefit. Because all the UEs are cooperative and individually rational with this assumption in place, the UEs that are in proximity to one another can form coalitions to provide connectivity and data sharing with each other. We assume that there is an arbitrator/coordinator who makes decisions with respect to the coalition structures, based on the energy profiles of all participating users. The simulation results show that utilization of this coalition clustering algorithm obtains energy efficiency and is potentially highly efficient in disaster situations as verified in our simulation results.

Algorithm 1 Coalition formation phases

Initialization
Cluster is formed with its proximity devices
One device act as UERCH $C_k = \{n_k\}$ and $\mid C_k \mid = 1$
All clusters are in search space $S = \{k; C_k \neq \emptyset\}$.

Phase 1 – Cluster and Coalition Candidate Search

Clusters are searched, based on its energy consumption per node. Find the highest energy consumption: $k = \arg\max_{i \in s} E_{C_i} / \mid C_i \mid$
Finding cluster C_j which when merged with C_k will consume lowest energy $j = \arg\min_{i \neq k} E_{C_j \cup C_k}$.

Phase 2 – Coalition formation

do
    if $E_{C_j \cup C_k} \leq E_{C_j} + E_{C_k}$ **then**
        Form a coalition between the members of clusters C_j and C_k
    else
        Work independently
    end if
    if Merger condition is satisfied **then**
        Set $\acute{C}_j = C_j \cup C_k$.
        n_j is the lowest energy consumption cluster head of the new coalition cluster.
    end if
    if Merger condition is not satisfied **then**
        keep cluster C_j and C_k separate
    end if
    Update the clusters, $C_j = \acute{C}_j$ and $C_k = \emptyset$.
while search space $S \neq \emptyset$

9.5 Results and Discussion

We consider an LTE coverage area with uniform UE distribution. A BS in coverage area transmitting at full power with 10 MHz of bandwidth is subdivided into 50 RBs of 12 subcarriers each. Channel and energy consumption parameters for simulation are taken from [26] and [27] and are shown in Table 9.1.

We use Matlab to analyze the impact of UE device variations, energy consumption of UEs, number of UEs, and number of clusters in field. A coalition efficiency method was used to carry out simulation which leads to significant energy saving as shown in the results. The scenario investigated in Figure 9.5 is based on one BS in the functional area transmitting at full power, while in disaster area the BS(s) are nonfunctional. In emergency conditions, users located in the nonfunctional area can take advantage from the D2D UE proximity services. Additionally, D2D relay

Table 9.1 Simluation parameters.

Parameters	Values	Parameters	Values
B (Bandwidth)	10 MHz	v	3.76
N_{RB}	50	T	1
S_T (Content size)	1Mbits	Rayleigh parameter	$E[a^2] = 1$
$P_{Tx,SR}$	1.425 J/s	Max. $UE_{TXpower}$	0.125 W
$P_{Rx,SR}$	0.925 J/s	$P_{Rx,LR}$	1.8 J/s

Source: Source: Ali et al. [18].

enhances the data throughput of edge-user(s) that can be used to link far away UEs with cellular coverage to the BS which extends the cellular coverage. We use clustering approach to reduce energy consumption and extend coverage area. A number of UEs form a cluster with one node acting as a UERCH, and the rest of the UEs in the cluster are called CMs.

In Figure 9.7, we have carried out simulation for 100 realizations of UE devices placement (uniformly distributed). The total energy consumption of UEs in non-clustering communication (without D2D) is plotted for varying number of UE devices. We can infer from Figure 9.7 that the energy consumption increases with the number of UE devices. Similarly, we have also plotted the total energy consumption of D2D by using coalition cluster formation algorithm for cluster formation and UERCH selection. In D2D communication, UE of each coalition

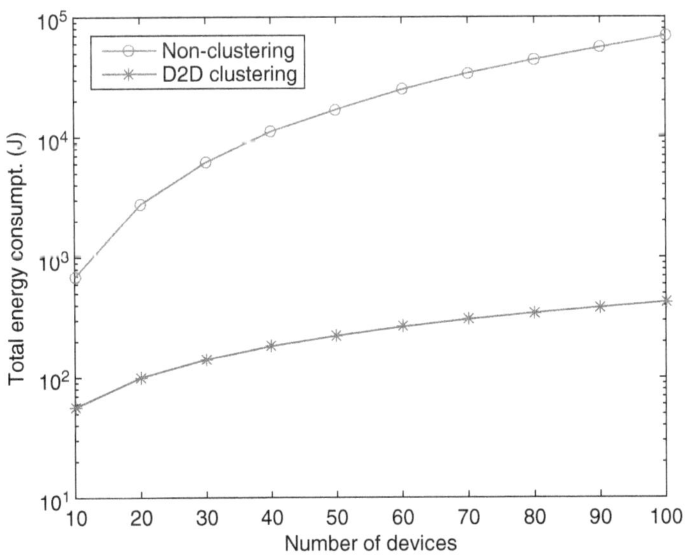

Figure 9.7 Comparison of total energy consumption of UEs devices in non-clustering- and D2D-based communication [18].

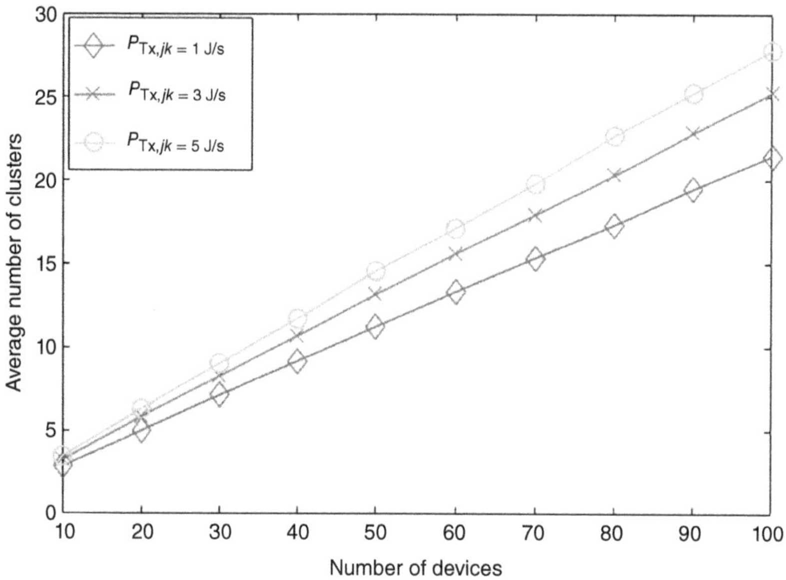

Figure 9.8 Average number of clusters [18].

communicates with their respective UERCH and then further with each URECH aggregate data from all CMs and transmit it via R to BS in functional area.

We compare the total energy consumption of UE devices in both non-clustering- and clustering-based D2D communication. We observe that the energy consumption in non-clustering scenario increases exponentially, whereas energy consumption increases almost linearly in D2D communication scenario.

We also investigated that impact of different number of UEs per cluster. Figure 9.8 is only for D2D scenario with cooperative clustering method by showing the variation in the power of each cluster and the average number of UEs in D2D network by implementing coalition cluster formation algorithm. The average number of clusters is increasing linearly, which means that it is easy to predict the number of clusters by the number of devices. When a UERCH transmits signals to its CM at high power, there are more number of clusters in the network, which means that each UERCH could have less UEs around, and this could also save energy and time to transmit data within clusters. Furthermore, in Figure 9.9, box plot provides a visualization of summary statistics for sample data. In our plot, we set outlier as a value that is more than 1.5 times the interquartile range away from the top or bottom of the box. The number of samples we have used to plot Figure 9.9 is 100 UE placements × 100 Rayleigh fading realizations. This is equivalent to 10^4 sample points. From the result, we can observe that the energy gain (in percentage) increases with the increase in the number of UE devices. We also note that the variation of energy value decreases with the increase in UE devices (i.e. light gray box).

The power of single subcarrier required for energy recovery as a function of number of devices for various EH efficiency values is shown in Figure 9.10. This power refers to the necessary single subcarrier power in such a way that harvested energy

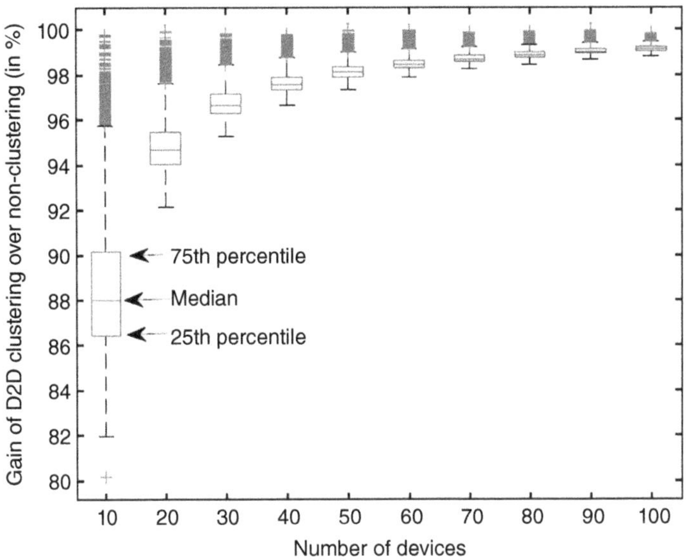

Figure 9.9 Gain of D2D clustering versus non-clustering box plot of energy gain (in percent) of UE devices in D2D clustering-based over non-clustering communication for varying number of UE devices [18].

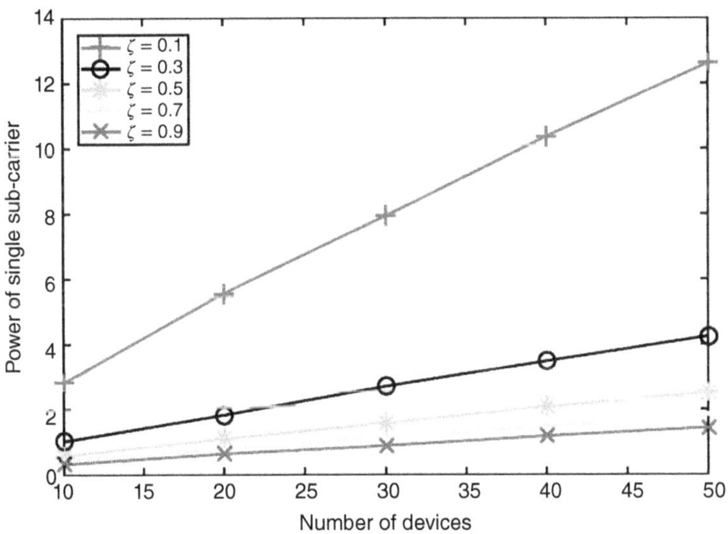

Figure 9.10 Power of single subcarrier [18].

from source is greater than or equal to the consumed energy required to transmit the signal from source to destination ($E \geq E_{C_k}$). This implies that the power of single subcarrier $p_n^{S,1}$ should be greater than or equal to a threshold. In fact, the represented curve corresponds to this threshold: values of $p_n^{S,1}$ greater than or equal to the curve satisfy the energy recovery equation (9.21), values of $p_n^{S,1}$ lower than the curve do

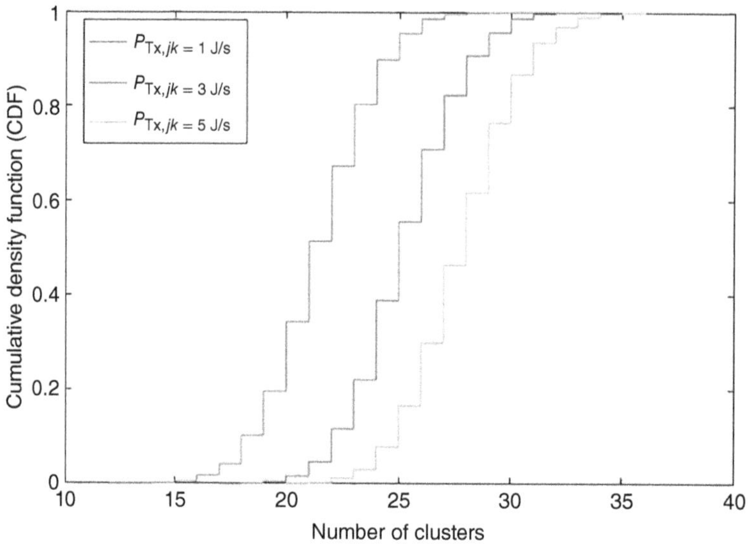

Figure 9.11 CDF for 100 UEs devices [18].

not satisfy the energy recovery equation, and hence harvested energy is not sufficient to transmit the signal from the source to the destination. We also notice that as the EH efficiency is getting larger, the represented threshold of single subcarrier power becomes smaller for the same number of devices. This means that when the energy is efficiently harvested, the power threshold is lowered and as a result, the system requires less energy for the subcarriers and important energy savings would be made.

We also evaluate cumulative distribution function (CDF) for clusters. Figure 9.11 shows the CDF of the number of clusters for 100 UE devices. The cluster formation algorithm applied on 100 UEs devices results in less than 25 clusters for 90% of time. We can also see that less than 22 clusters are obtained only for 30% of time during small power (1 J/s). As we are able to increase the level of power to be received, the formation of clusters will take less time, which accordingly reflects the efficiency of the proposed system. It means that by using this approach, we will be able to initiate communications in the disaster area more rapidly. The link between relay and the UER is therefore of great significance, especially in our proposed PS networks model. Furthermore, deploying clustering and EH techniques to keep the nodes alive and passing information further in nonfunctional area is a robust physical network design for survivable networks.

The Figure 9.12 represents the average number of clusters in D2D communications and power used for transmission (power drained from the battery of node n_k to transmit data to node n_j) for different sizes of transmitted data contents. The number of clusters increases as $P_{Tx,jk}$ increases. This means that when a node n_k uses more power to transmit data, it is more likely that the number of formed clusters increases.

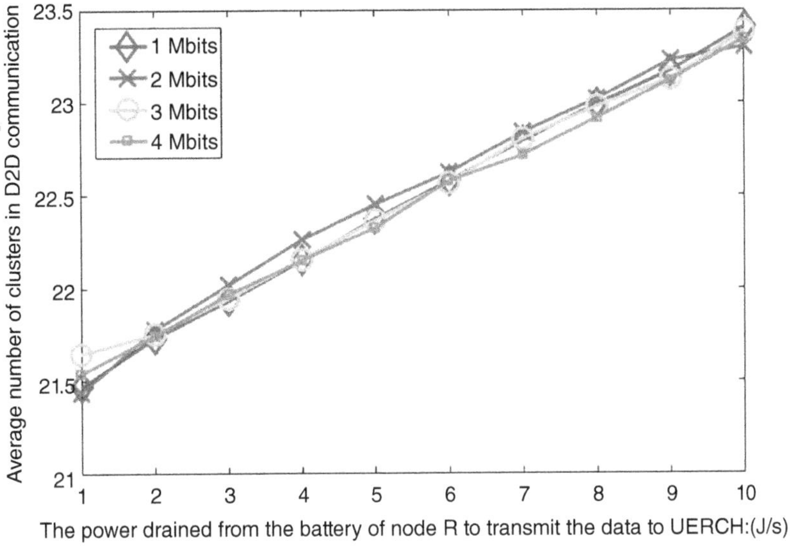

Figure 9.12 The power drained from the battery of relay to transmit the data to UERCH [18].

Moreover, clusters are formed to reduce energy consumption which confirms the results found in Figures 9.8 and 9.9.

9.6 Summary

This chapter has highlighted the use of D2D communications in PSN. We have briefly introduced the concept of RF-based EH and its applications as well as its potential benefits in this area. We have presented a new design of robust network in disaster and emergency situations. The aim was to provide the optimal communication route for networks in disaster areas, which minimizes the end-to-end disconnection and enables the connection from functional area to nonfunctional area. In our proposed approach, we have considered a UER that uses RF EH via base station for D2D communications. Clustering technique is integrated with D2D communications into cellular networks, such that communication services can be maintained when the cellular infrastructure becomes partially dysfunctional. Moreover, a novel concept of power transfer in D2D clustering with UERCH is proposed to provide a new framework to handle critical and emergency situations. The proposed approach was shown to provide significant energy savings for both UE nodes and clustering heads to survive in critical situations like disasters. In addition, our proposed method adds a new step to the provisioning phase for network survivability against network failure and can be executed in an incremental fashion. It can also be combined with other protection and restoration methods to enhance network robustness post-disaster to provide better link connection.

References

1 Kumbhar, A., Koohifar, F., Fuvenc, I., and Mueller, B. (2017). A survey on legacy and emerging technologies for public safety communications. *IEEE Communication Surveys and Tutorials* 19: 1: 97–124.

2 Ali, K., Nguyen, H.X., Vien, Q.T., and Shah, P. (2015). Disaster management communication networks: challenges and architecture design. *Pervasive Computing and Communication Workshops (PerCom)*, USA, Mar 2015.

3 Sudevalayam, S. and Kulkarni, P. (2011). Energy harvesting sensor nodes: survey and implications. *IEEE Communication Surveys and Tutorials* 13 (3): 443–461.

4 Lu, X., Wang, P., Niyato, D. et al. (2014). Wireless networks with RF energy harvesting: A Contemporary Survey. *IEEE Communication Surveys and Tutorials* 17: 757–789.

5 Zhou, X., Zhang, R., and Ho, C.K. (2013). Wireless information and power transfer: architecture design and rate-energy tradeoff. *IEEE Transactions on Communications* 61 (11): 4754–4767.

6 Ali, K., Nguyen, H.X., Shah, P. et al. (2016). Architecture for public safety network using D2D communication. *Proceedings of IEEE Wireless Communications and Networking Conference (WCNC)*, Doha, April 2016.

7 Statt, N. (2016). Facebook launches crisis response hub to help users during disasters and attacks, The Verge. https://www.theverge.com/2017/9/14/16305206/facebook-crisis-response-center-safety-check-update.

8 Asadi, A., Wang, Q., and Mancuso, V. (2014). A survey on device-to-device communication in cellular networks. *IEEE Communication Surveys and Tutorials* 16 (4): 1801–1819.

9 Nishiyama, H., Ito, M., and Kato, N. (2014). Relay-by-smartphone: realizing multihop device-to-device communications. *IEEE Communications Magazine* 52 (4): 56–65.

10 Hunukumbure, M., Moulsley, T., Oyawoye, A. et al. (2013). D2D for energy efficient communications in disaster and emergency situations. *in Software, Telecommunications and Computer Networks (SoftCOM), 2013 21st International Conference on. IEEE.*

11 Chalise, B.K., Zhang, Y.D., and Amin, M.G. (2012). Energy harvesting in an OSTBC based amplify-and-forward MIMO relay system. *Proceedings of IEEE International Conference on Acoustics, Speech and Signal Processing (ICASSP)*, Kyoto, March 2012.

12 Lee, C.H. (2014). Wireless information and power transfer for communication recovery in disaster areas. *Proceedings of IEEE 15th International Symposium on World of Wireless, Mobile and Multimedia Networks (WoWMoM), Sydney, June 2014.*

13 Li, M., Nishiyama, H., Kato, N. et al. (2015). On the energy-efficient of throughput-based scheme using renewable energy for wireless MESH networks in disaster area. *IEEE Transactions on Emerging Topics in Computing* 3 (3): 420–431.

14 Nahas, M., Ghantous, M., Al haj, K., and Assaf, B. (2014). For better energy consumption and management in future cellular networks. *Proceedings of International Conference on Renewable Energies for Developing Countries (REDEC), Beirut, Noveber 2014.*

15 Sakano, T., Kotabe, S., Komukai, T. et al. (2016). Bringing movable and deployable networks to disaster areas: development and field test of MDRU. *IEEE Network* 30 (1): 86–91.

16 Altay, C., Bozdemir, N.Z., and Camcıoglu, E. (2016). Standalone eNode-B design with integrated virtual EPC in public safety networks. *Proceedings of IEEE/IFIP Network Operations and Management Symposium (NOMS)*, Istanbul, April 2016.

17 Lei, X., Fuja, T.E., and Costello, D.J. (2010). Mobile relaying: coverage extension and throughput enhancement. *IEEE Transactions on Communications* 58 (9): 2709–2717.

18 Ali, K., Nguyen, H.X., Vien, Q. et al. (2018). Disaster management using D2D communication with power transfer and clustering techniques. *IEEE Access* 6: 14643–14654.

19 Huang, G., Zhang, Q., and Qin, J. (2015). Joint time switching and power allocation for multicarrier decode-and-forward relay networks with SWIPT *IEEE Signal Processing Letter* 22 (12): 2284–2288.

20 Nasir, A.A., Zhou, X., Durrani, S., and Kennedy, R.A. (2013). Relaying protocols for wireless energy harvesting and information processing. *IEEE Transactions on Wireless Communications* 12 (7): 3622–3636.

21 Ding, Z., Krikidis, I., Sharif, B., and Poor, H.V. (2014). Wireless information and power transfer in cooperative networks with spatially random relays. *IEEE Transactions on Wireless Communications* 13 (8): 4440–4453.

22 Yang, H.H., Lee, J., and Quek, T.Q.S. (2014). Green device-to-device communication with harvesting energy in cellular networks. *Proceedings of the 6th International Conference on Wireless Communications and Signal Processing (WCSP), Hefei, October 2014.*

23 Ali, K., Nguyen, H.X., Shah, P., and Vien, Q.T. (2016). Energy efficient and scalable D2D architecture design for public safety network. *Proceedings of Advanced Communication Systems and Information Security (ACOSIS)*, Marrakesh, October 2016.

24 Li, Z. (2012). Performance Analysis of Network Assisted Neighbor Discovery Algorithms. *Tech. Rep. XR-EE-RT 2012:026*. Stockholm, Sweden: School Elect. Eng., Royal Inst. Technol. http://urn.kb.se/resolve?urn=urn:nbn:se:kth:diva-117696 (accessed 08 October 2021).

25 Chatterjee, M., Das, S.K., and Turgut, D. (2002). WCA: A weighted clustering algorithm for mobile ad hoc networks. *Cluster Computing* 5 (2): 193–204.

26 Yaacoub, E. and Kubbar, O. (2012). Energy-efficient device-to-device communications in LTE public safety networks. *Globecom Workshops 2012 (International Workshop on Green Internet of Things)*, Anaheim, CA, December 2012.

27 Khan, S. and Mauri, J.L. (2013). Interplay between cooperative D2D communication and green LTE cellular networks. In: *Green Networking and Commuications: ICT for Sustainability* (eds. S. Khan, J. Lloret Mauri), Chapter 7, 148–155. New York: CRC Press.

10

Road Ahead for D2D Communications

Masood Ur Rehman[1], Ghazanfar Ali Safdar[2], and Mohammad Asad Rehman Chaudhry[3]

[1]*James Watt School of Engineering, University of Glasgow, Glasgow, United Kingdom*
[2]*School of Computer Science and Technology, University of Bedfordshire, Luton, United Kingdom*
[3]*Soptimizer, Canada*

The global market of Internet-of-Things (IoT) devices was valued USD 31.5 billion in 2016. It is expected to reach around 158 billion in 2024 experiencing a compound annual growth rate (CAGR) of 23% [1]. 5G is expected to be the strongest enabler in the expansion of the IoT due to unparalleled speed, large bandwidth, massive capacity, low latency, towering scalability, and high reliability. The 5G network has the capability to support a huge number of static and mobile IoT devices, having a diverse range of speed, bandwidth, and quality of service requirements. Device-to-device (D2D) is a key enabling technology for the 5G IoT, where the global number of D2D connections specifically is expected to reach 14.7 billion devices by 2023 [2].

This mammoth growth in the market is driven by factors such as increasing population, sharply growing use of mobile devices, upsurge in the integration of virtual reality, augmented reality and artificial intelligence, expanding subscriptions of over-the-top (OTT) platforms, intensive penetration of internet users, and rising audience of e-sports. This market growth would, however, be contested by a number of challenges such as security and privacy issues, massive power consumption, and high operational and infrastructural cost [3].

10.1 Future Prospects and Challenges

This book has introduced new approaches to mitigate interference in D2D communications by means of novel algorithms and system-level proposals. The importance of D2D in 5G networks cannot be overemphasized. Although D2D can retrieve operating parameters, success of current and future D2D applications and hence the growth and innovation in the implementation of 5G depends on efficient solutions to the technological factors it faces today. Some of the key prospects and challenges are discussed in the following sections.

Interference Mitigation in Device-to-Device Communications, First Edition.
Edited by Masood Ur Rehman, Ghazanfar Ali Safdar, and Mohammad Asad Rehman Chaudhry.

10.1.1 Spectrum Sharing and Coexistence

The emergence of modern wireless technologies has resulted in a tremendous increase in wireless data traffic. One straight approach to accommodate this huge data is to find more spectrum by which is a stringent task due to the spectrum availability and standardization issues. 5G and beyond networks aim to solve this problem by enhancing the spectrum efficiency employing advanced spectrum sharing techniques such as cognitive radio, in-band full-duplex, nonorthogonal multiple access (NOMA), long-term evolution on unlicensed spectrum (LTE-U), and D2D communication [4]. The basic idea of spectrum sharing is to allow transceivers to use idle or underutilized spectrum band temporally and geographically. Spectrum sharing is indeed a promising strategy to tackle the imbalance between limited spectrum resource and unprecedent traffic demands. In D2D communication, the spectrum can be shared between D2D and cellular users via two methods, namely, in-band (licensed) and out-band (unlicensed) communication [5]. Cognitive radio technologies make it possible to share this spectrum by using radio environmental awareness techniques and interference management that allow multiple systems to occupy the same spectrum. Meanwhile, more types of cognition can be explored to combine the promising authorized/licensed shared access and coprimary sharing, for instance, traffic cognition and spectrum resource reservation.

Consequently, future communication systems are believed to have ultra-dense node deployment. These networks are, therefore, required to support heterogeneous technologies for compatibility of various user's types (such as smartphones, tablets), network types (such as cloud and edge computing, IEEE 802.11, IoT), and Quality of Service types (such as varying speed, delay, and throughput for specific applications) [6]. Coexistence of D2D communication with Mobile Edge Computing (MEC), Software-defined Networking (SDN), and Network Function Virtualization (NFV) has exhibited substantial potential to improve the throughput of the system even in imperfect channel conditions [7–9]. D2D delivers this improvement by offering reuse gain (same resources can be used by both cellular and D2D users), proximity gain (based on the distance between D2D devices to provide increased bit rate, low power consumption, and less delay), and hop gain (D2D devices can use a single link instead of uplink and downlink). Nevertheless, the integration of D2D in the network requires a substantial modification in both core and access networks to support the smart services in future innovative infrastructures [10].

10.1.2 Standardization

As a diverse range of devices needs to be connected in D2D, the necessity to coordinate with these different devices has increased. Various devices use different hardware, run over different platforms, and are manufactured by different vendors. This incompatibility among devices, sensors, and even interfaces of different applications cause the interoperability challenge in the dearth global standards or policies. Several associations and organizations are working toward resolution of this issue. The existing interoperability standards, such as MTConnect, Ethernet for Control Automation Technology (EtherCAT), MCS-DCS Interface Standardization (MDIS), Master Control System (MCS), Distributed Control System (DCS), and Interface Standardization, promote data interchange across heterogeneous domains and industries [11]. Companies are currently using custom Application Programming Interfaces (APIs) and solutions. A common protocol and communication standard is required especially in the decentralized D2D network environment to enable secure interaction and data transmission among various devices [12].

10.1.3 Secure Communication

User equipment (UE) communicates through radio network links and core network in conventional cellular communication. On the other hand, D2D can handle connection and data traffic without the core network enabling the UE to communicate directly with other D2D-enabled UEs in proximity using the D2D link. This brings in added security risks on top of the security threats due to the huge number of connected devices in 5G IoT making security assurance one of the major challenges for D2D [13].

The D2D communication can be threatened from node impersonation, message modification, eavesdropping, masquerading, and man-in-the-middle attacks. Elliptic Curve Cryptography, AES-128, and access points (APs) are some of the techniques that are traditionally used for protecting privacy and security in such scenarios [14]. These techniques are, however, not suitable for a distributed, heterogenous Opportunistic, and self-organizing network like D2D requiring a self-reliant anonymity scheme independent of a centralized single trusted center (i.e. base station (BS)). Enhanced authentication and key agreement mechanisms are, therefore, required to secure D2D communication [15].

The security in D2D can be ensured if closed access is applied to devices (Figure 10.1) providing each device with a list of reliable peers, such as the users in the close vicinity, office to whom they are familiar with and the users that have been legitimated through a reliable party like an association. These users can unswervingly communicate with each other, sustaining a level of discretion, whereas the devices not on this list need to use the macro cell level to communicate with it. Open access where each device can turn in to relay for other devices deprived of any limits can severely daunt the security in D2D. Data sharing using the blockchain technology can be a worthwhile option where authentication is obtained by deploying APs verifying the transactions of end users followed by a transaction relaying phase and the block verification phase [16]. Evaluation of trustworthiness of a user based on the "social" reputation considering the brand, type of friendship with other nodes, and computational capabilities is also being explored to ensure security in D2D [17]. Security and reliability are, however, an open research problem in D2D communications.

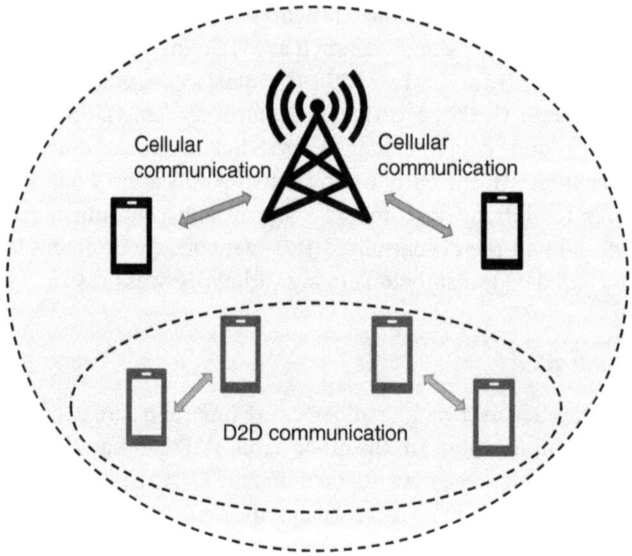

Figure 10.1 Closed access 5G and D2D communication.

10.1.4 Energy Consumption and Energy Harvesting

Sustainability of the D2D communications depends upon the battery constraints of the UEs. D2D is considered as a key enabler of 5G IoT that has densified the communication network in order to increase the coverage and capacity of the system. The resulting surge in power consumption along with the short battery life of UEs and the requirements for better quality of service has made energy efficiency a challenging issue in D2D [18, 19].

The power consumption in today's mobile systems is dominated by the network with up to 90% as compared to only 10% for mobile terminals [20]. Though direct link in D2D communication can increase network energy efficiency by offloading some energy from the BS, battery powered D2D devices must look for ways to improve on the energy consumption.

Deep reinforcement learning (DRL) and a two-way amplify-and-forward relay can be useful to devise effective power control in D2D environments [21, 22]. RF energy harvesting also appears as a possible solution to optimize energy management in D2D technology [23, 24]. D2D communications may also use wireless power transfer and optimization of sleep mode switching in order to have a battery-less operation. Furthermore, Simultaneous Wireless Information and Power Transfer (SWIPT) is also a promising candidate for power efficient D2D communication in which same electromagnetic field can be used both for information and energy transfer to the end users and/or devices [25–28]. The combined receiver design for SWIPT can separate information and energy circuits, as illustrated in Figure 10.2, or it can integrate both in the same circuit, as shown in Figure 10.3.

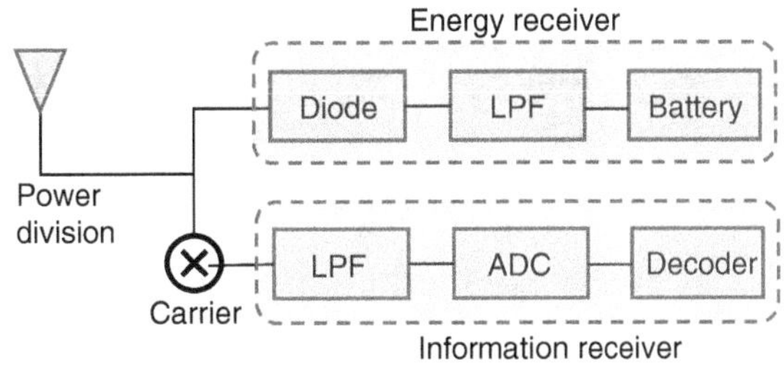

Figure 10.2 Separated information and energy receiver.

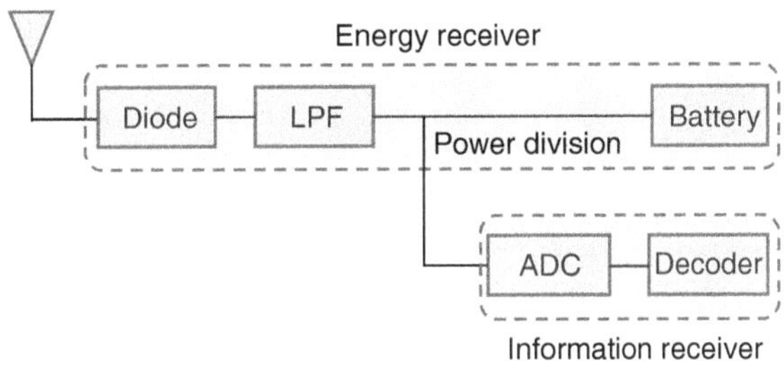

Figure 10.3 Integrated information and energy receiver.

10.1.5 Interference Management

D2D users, depending upon the mode (i.e. uplink or downlink), share the same licensed spectrum with cellular users, and users prefer the sharing mode to increase the spectral efficiency. Due to the introduction of D2D communication to the cellular network, the cellular architecture generally includes two tiers; the conventional macrocell layer, which involves the communication between BS and device; and the device tier, which involves D2D communication. By reusing wireless resources of the cellular network and coexistence with cellular users, D2D communications inevitably generate interference [18, 21]. In the two-tiered network architecture, two types of interferences are introduced: co-tier and cross-tier. Co-tier interference is produced between network elements that belong to the same tier in the network, such as interference between a D2D users and another D2D user in the same tier. Cross-tier interference is produced between network elements that belong to different tiers, i.e. interference between D2D user and cellular user, as shown in Figure 10.4.

Interference management between cellular users and D2D users is a critical issue. In general, interference management schemes can be classified into centralized, distributed, and semi-distributed approaches. In the centralized approach, BS fully manages interference between cellular and D2D user. In a distributed scheme, the operation of interference management does not require a central entity and is performed autonomously by D2D users themselves. Using the advantages and trade-offs of centralized and distributed schemes, semi-distributed or hybrid interference management scheme can be used. Based on the type of interference, several interference management schemes are proposed such as spectrum splitting, power control, mobile selection, radio resource allocation, MIMO, Beamforming, and coding [29–31]. We have discussed some of these schemes in different D2D application scenarios in detail in the earlier chapters.

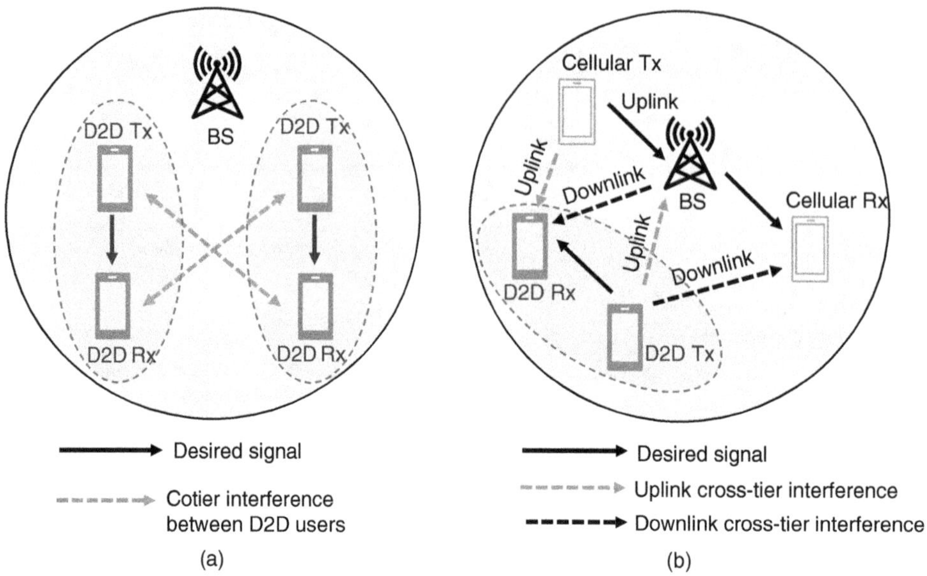

Figure 10.4 (a) Co-tier interference (b) cross-tier interference.

5G multitier heterogeneous systems involving macrocell, picocell, femtocell, and D2D pairs with dense node deployment would increase computational complexity exhibiting a complex interference management problem as different tiers produce different levels of interference due to varying access restrictions [14, 18]. It would require improvements in existing interference management techniques to enhance the spectrum efficiency, such as relay aided D2D at cell edges if the deep fading links are observed, different power allocation schemes for uplink and downlink, and cooperative links between macrocells and small cells.

10.1.6 Resource Allocation

Efficient and intelligent allocation of scarce spectrum resources is vital to minimize the interference issue and improve the overall performance of the D2D communications in terms of throughput, energy efficiency, channel utilization, capacity, reliability, latency, and quality of service [32, 33]. The resource allocation approaches can be centralized where optimization procedures are executed by the BS. D2D users provide information on the local channel quality measurements to the BS rendering it to have a complete knowledge of the Channel State Information (CSI) in the network. However, when the CSI is not known at the BS, the centralized resource allocation technique is inefficient leading to the need of a distributed resource allocation procedure. The distributed resource allocation methods are device centric where the D2D devices themselves sense and utilize the available spectrum. The BS is involved only for freezing the spectrum requested by the D2D devices. The D2D devices share the local channel information with the near neighboring devices. The main disadvantage of using distributed resource allocation is that the number of message exchanges

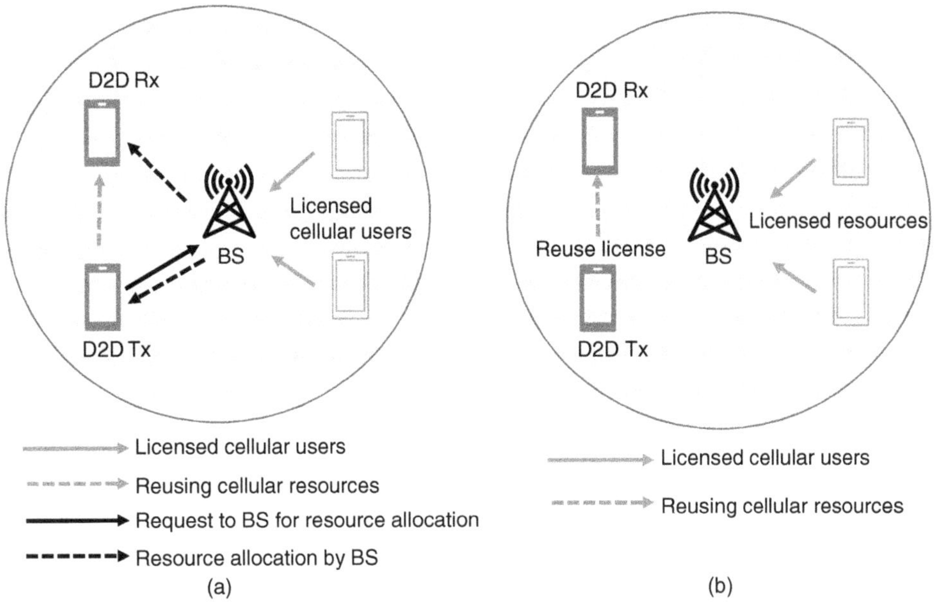

Figure 10.5 (a) Centralized resource allocation (b) distributed resource allocation.

between the devices is greater. Concept of centralized and distributed D2D resource allocation is illustrated in Figure 10.5.

Various resource allocation techniques are being used in D2D communication such as Heuristic algorithm, Lagrangian duality, Evolutionary, Steepest descent, Game Theory, Graph Theory, and Fuzzy logic [34–36]. Densification of the D2D nodes in 5G IoT will manifold increase the complexity of resource allocation, and hybrid/joint optimization methods would be required to ensure sustainability of the D2D communications.

10.1.7 Device Discovery

Efficient discovery of the nodes in proximity is fundamental to the D2D communications. The beacon signals exchanged between the devices contain the information such as device ID and CSI to establish the plausibility of being paired together [37]. A device discovery procedure can also be divided into centralized and distributed device discovery [38]. A central entity (e.g. an access point or a BS) assists the devices in discovering one another in the centralized approach, while the devices are permitted to discover other devices without including the BS in the distributed approach. Use of random back-off procedure, sequential estimation, elliptic-curve Diffie–Hellman method, and public safety full duplexing are some of the approaches being employed to perform efficient device discovery [14, 39–41].

Sustainability of the 5G D2D networks is strongly linked to the quick device discovery process to uphold the battery constraints and avoid discontinuations due to node mobility. A fast and dynamic scheme capable of exploiting network assisted and UE initiated device discovery conforming to the circumstantial changes and supporting a multicell environment with minimal signaling overhead would be important for future D2D communications.

10.1.8 Handover

Handover is an important issue in cellular networks due to the mobility of UEs. Smooth handover is necessitous to provide pervasive communication particularly in futuristic multicell environments. The handover decision is mainly based on UE's report on signal strength and quality, such as reference signal received power (RSRP), received signal strength indicator (RSSI), and reference signal received quality (RSRQ) [18, 42].

Although the data can be directly transmitted among D2D devices in D2D communication, the link between D2D devices and BS is still needed for controlling the transmit power of the device and resource allocation. D2D handover management has two issues that need efficient solutions. First, it is possible that both devices of a D2D pair are connected to different BSs at the cell boundary leading to additional latency because of the signal exchanges between the two BSs. Therefore, a mechanism should be devised to allow a D2D pair to be associated with the same BS in order to reduce signaling complexity and overhead among BSs and D2D devices. Second, the movement of UEs can generate shadow fading in a real environment resulting in unnecessary handovers between two BSs within a short interval. This may happen when a D2D user moves into the overlapped coverage region of multiple BSs or is shielded by buildings, vegetation, or other obstructions. The consequential also generates additional latency and degrades the system performance due to the reallocation of resources to the D2D users [43].

10.1.9 Network Slicing

Network slicing is a novel technique that permits innovative models for corporate sector and provides a way to achieve the dynamic network requirements such as increased network size, increased traffic volume, constantly changing networks, and increased network complexity. This will enable the network operators to slice the single conventional network into multiple logical networks. The network slicing provides an innovation sandbox with a level of isolation granting network operators the ability to develop and deploy novel solutions without impacting overall network functions. It can be a useful tool for 5G system developers enabling new user experiences through simplified operations, reduced cost, reduced time to market, efficient and optimized use of network resources with better insight and planning, use of intelligent automation to gain an advantage in evolving and highly competitive landscape, quick fulfilment of customer's demands, and improve on reliability and scalability of the network [44]. For D2D communication, network slicing has been done using SDN and virtualized networks where the SDN controller or slice controller is responsible to direct the D2D connection for a successful session. Though aimed at minimizing the delay and hardware expenses as compared to current wireless networks, network slicing in D2D communications is still in its infancy and need structured efforts to have a meaningful method in place [45].

10.1.10 D2D in Vehicular Communications

Vehicle-to-vehicle (V2V) communications have attracted great interest due to the potential of improving traffic safety, reducing energy consumption, and enabling new services related to intelligent transportation systems. These applications have a strong localized nature, i.e. requiring cooperation between vehicles in close proximity. Furthermore, V2V has strict requirements of real-time operation, reliability, and access. D2D communication has been considered as an enabler for V2V communication [46]. The localized nature of V2V services is the motivating idea for D2D communication. Moreover, the low latency requirement of V2V applications fits the hop gain of D2D transmissions. Besides, V2V's requirement on high reliability is consistent with the proximity gain provided by D2D links. Guaranteeing the required latency, capacity, reliability, network utilization, and intra-cell interference is challenging for V2V services. The instantaneous changes in the channel conditions due to mobility of vehicles make D2D resource management techniques assuming perfectly known CSI unsuitable for V2V communications. Hence, radio resource management is a key design aspect and open challenge in D2D-based V2V communication [47, 48].

10.1.11 D2D in Disaster Management

Communication infrastructure is vital for the organization of rescue services in the aftermath of a natural disaster events such as earthquake, landslide, flood, and storm surge. However, the typical communication network infrastructure often receives the first hit in the disaster region. The dysfunctional communication network due to damaged communication infrastructure, loss of energy/power or low reliability, and availability of network in the event of a natural calamity can severely dent the rescue operations. In such circumstances, D2D communications can employ its proximity service to enable the connection from functional area to nonfunctional area providing structureless communications for emergency and safety information. D2D facilitates extension of cellular network coverage by forming multihop UE-to-UE relaying that allows an out-of-coverage UE to link to the BS and receive emergency information [49]. Use of drones, energy-efficient clustering, and routing are some of the aspects being exploited for D2D-based disaster management systems [50–52]. Though promising, several issues such as limited battery, range between estranged devices, bandwidth, disconnectivity, network overload, data redundancy, communication overhead, network lifetime, lack of information, and data integration need satisfactory answers to exploit the full potential of D2D in disaster management.

10.1.12 D2D at Millimeter Wave Frequencies

Millimeter wave (mmWave) frequencies of 30–300 GHz are a front-running candidate for 5G D2D systems due to their inherent features of device miniaturization, low interference, and high data rates [53]. Short range and highly directional coverage through beamforming make them highly suitable for D2D proximity services. However, high propagation losses, user mobility and obstruction from human users, buildings, vehicles, and vegetation can severely affect these benefits [54, 55]. Moreover, fundamental changes in the device hardware would also be required [56]. Therefore, efforts are needed to improve the hardware interface, standardize specifications, devise effective solutions to minimize frequent handovers, and interference for a moving node, characterize the human user effects and develop traffic offloading methods for coexistence with cellular networks to fully appreciate the advantages of D2D communication at mmWave frequencies.

10.1.13 D2D and Social Networks

Relationship tracing of D2D nodes through trusted social networks can help building secure connections, sustainable network, and enhanced information exchange. Pairing of devices with social ties, similar interests, and common content sharing can be realized based on this information [57, 58]. This leads to efficient device discovery by avoiding random beaconing as beacon rates can be adjusted according to strengths of ties. Taking social tie information into account in resource allocation improves privacy and security as strength of a tie reflects trustfulness of two devices [59]. It can also help in efficient spectrum utilization by offloading traffic from the BS to D2D network as the BS would transmit a content to one of the D2D UE who would then relay it to the devices carrying similar interests [18, 60]. However, steps need to be taken to ensure reliability and rule out impersonation. Also, link dynamicity should be provided to allow self-adjustment mechanism in the wake of relocation of a node.

10.1.14 D2D and Visible Light Communication (VLC)

RF technology suffers from the capacity crunch, interference, and other constraints that affect the performance of the communication system. Visible light communication (VLC) has been identified to substantially overcome these issues [61, 62]. VLC

Table 10.1 RF and VLC technologies in comparison.

Parameters	RF	VLC
Spectrum band	~3–300 GHz	~400–780 THz
Spectrum allocation	Licensed	Unlicensed
Power consumption	Medium	Low
Complication	High	Low
Multipath fading	Yes	No
Coverage area	Wide	Bounded
Electromagnetic interference	Yes	No

has high bandwidth reaching up to Terahertz (THz), in addition to low power, license free, and immunity to interference [63]. Table 10.1 compares some key features of RF and VLC technologies.

Due to the short range, VLC is more suitable for indoor environments. Employing hybrid RF-VLC approach in D2D communication has shown the ability to improve capacity and achieve high data rates [64]. Therefore, VLC-based D2D can potentially be a venue for more structured research and development efforts in near future.

References

1 Research and Markets (2018). IoT Devices Market by Device Type: Global Industry Analysis, Size, Share, Growth, Trends, and Forecast 2016-2024. https://www.researchandmarkets.com/reports/4531902/iot-devices-market-by-device-type-global (accessed 21 October 2021).

2 Research and Markets (2021). Global 5G Smartphone Market: Insights & Forecast with Potential Impact of COVID-19 (*2021*-2025). https://www.researchandmarkets.com/reports/5308210/global-5g-smartphone-market-insights-and#rela0-4856120 (accessed 21 October 2021).

3 Markets and Markets (2021). 5G IoT Market by Component. https://www.marketsandmarkets.com/Market-Reports/5g-iot-market-164027845.html (accessed 21 October 2021).

4 Safdar, G.A., Ur-Rehman, M., Muhammed, M. et al. (2016). Interference mitigation in D2D communication underlaying LTE-A network. *IEEE Access* 4: 7967–7987.

5 Ur-Rehman, M. and Safdar, G.A. (editors) (2018). LTE communications and networks: Femtocells and antenna considerations. *Wiley & Sons (UK).*

6 Kanwal, K., Safdar, G.A., and Ur-Rehman, M. (2018). Call blocking and outage probability in energy-efficient LTE networks. *Transactions on Emerging Telecommunications Technologies* 29 (10).

7 Moubayed, A., Shami, A., and Lutfiyya, H. (2015). Wireless resource virtualization with device-to-device communication underlaying LTE network. *IEEE Transactions on Broadcasting* 61 (4): 734–740.

8 Cai, Y., Yu, F.R., Liang, C. et al. (2016). Software-defined device-to-device (D2D) communications in virtual wireless networks with imperfect network state information (NSI). *IEEE Transactions on Vehicular Technology* 65 (9): 7349–7360.

9 Wang, K., Li, H., Richard Yu, F., and Wei, W. (2016). Virtual resource allocation in software-defined information-centric cellular networks with device-to-device communications and imperfect CSI. *IEEE Transactions on Vehicular Technology* 65 (12): 10011–10021.

10 Nadeem, L., Azam, M.A., Amin, Y. et al. (2021). Integration of D2D, network slicing, and MEC in 5G cellular networks: survey and challenges. *IEEE Access* 9: 37590–37612.

11 Kar, U.N. and Sanyal, D.K. (2020). A critical review of 3GPP standardization of device-to-device communication in cellular networks. *SN Computer Science* 1 (37).

12 Höyhtyä, M., Apilo, O., and Lasanen, M. (2018). Review of latest advances in 3GPP standardization: D2D communication in 5G systems and its energy consumption models. *Future Internet* 10 (1): 3.

13 Suraci, C., Pizzi, S., Garompolo, D. et al. (2021). Trusted and secured D2D-aided communications in 5G networks. *Ad Hoc Networks* 114.

14 Adnan, M.H. and Zukarnain, Z.A. (1762). Device-to-device communication in 5G environment: issues, solutions, and challenges. *Symmetry* 12: 2020.

15 Hamoud, O.N., Kenaza, T., and Challal, Y. (2018). Security in device-to-device communications (D2D): a survey. *IET Networks* 7: 14–22.

16 Jiang, L., Xie, S., Maharjan, S., and Zhang, Y. (2020). Joint transaction relaying and block verification optimization for blockchain empowered D2D communication. *IEEE Transactions on Vehicular Technolology* 69: 828–841.

17 Pizzi, S., Suraci, C., Iera, A. et al. (2020). A sidelink-aided approach for secure multicast service delivery: from human-oriented multimedia traffic to machine type communications. *IEEE Transactions on Broadcasting* 67 (1): 313–323.

18 Ansari, R.I., Chrysostomou, C., Hassan, S.A. et al. (2018). 5G D2D networks: techniques, challenges, and future prospects. *IEEE Systems Journal* 12 (4): 3970–3984.

19 Arabas, P. (2018). Energy aware data centers and networks: a survey. *Journal of Telecommunications and Information Technology* (4): 26–36.

20 Kanwal, K., Safdar, G.A., Ur-Rehman, M., and Yang, X. (2017). Energy management in LTE networks. *IEEE Access* 5: 4264–4284.

21 Wang, D., Qin, H., Song, B. et al. (2021). Joint resource allocation and power control for D2D communication with deep reinforcement learning in MCC. *Physical Communication* 45.

22 Han, L., Zhou, R., Li, Y. et al. (2020). Power control for two-way AF relay assisted D2D communications underlaying cellular networks. *IEEE Access* 8: 151968–151975.

23 Naqvi, S.A.R., Pervaiz, H., Hassan, S.A. et al. (2018). Energy-aware radio resource management in D2D-enabled multi-tier HetNets. *IEEE Access* 6: 16610–16622.

24 Lu, X., Wang, P., Niyato, D. et al. (2015). Wireless networks with RF energy harvesting: a contemporary survey. *IEEE Communication Surveys and Tutorials* 17 (2): 757–789.

25 Rajaram, A., Khan, R., Tharranetharan, S. et al. (2019). Novel SWIPT schemes for 5G wireless networks. *Sensors* 19: 1169.

26 Ge, X., Jia, H., Zhong, Y. et al. (2019). Energy efficient optimization of wireless-powered 5G full duplex cellular networks: a mean field game approach. *IEEE Transactions on Green Communications and Networking* 3 (2): 455–467.

27 Alemaishat, S., Saraereh, O.A., Khan, I., and Choi, B.J. (2019). An efficient resource allocation algorithm for D2D communications based on NOMA. *IEEE Access* 7: 120238–120247.

28 Hassani, S.E., Hassani, H.E., and Boutammachte, N. (2019). Overview on 5G radio frequency energy harvesting. *Advances in Science, Technology and Engineering Systems Journal* 4 (4): 328–346.

29 Noura, M. and Nordin, R. (2016). A survey on interference management for device-to-device (D2D) communication and its challenges in 5G networks. *Journal of Network and Computer Applications* 71: 130–150.

30 Shamaei, S., Bayat, S., and Hemmatyar, A.M.A. (2018). Interference management in D2D-enabled heterogeneous cellular networks using matching theory. *IEEE Transactions on Mobile Computing* 18 (9): 2091–2102.

31 Chen, L., Deng, N., and Wei, H. (2020). Performance analysis of D2D and cellular coexisting networks with interference management. *IEEE Access* 8: 82747–82759.

32 Hussein, H.H., Elsayed, H.A., and Abd El-kader, S.M. (2020). Intensive benchmarking of D2D communication over 5G cellular networks: prototype, integrated features, challenges, and main applications. *Wireless Networks* 26: 3183–3202.

33 Ashfaq, K., Safdar, G.A., and Ur-Rehman, M. (2021). Comparative analysis of scheduling algorithms for radio resource allocation in future communication networks. *PeerJ Computer Science* 7: e546.

34 Han, Q., Yang, B., and Wang, X. (2019). Online and robust resource allocation for D2D communications assisted by Green relays. *IET Communications* 13 (20): 3547–3557.

35 Huang, J., Xing, C.-C., Qian, Y., and Haas, Z.J. (2017). Resource allocation for multicell device-to-device communications underlaying 5G networks: a game-theoretic mechanism with incomplete information. *IEEE Transactions on Vehicular Technology* 67 (3): 2557–2570.

36 Zhang, Y., Zheng, J., Lu, P.-S., and Sun, C. (2016). Interference graph construction for cellular D2D communications. *IEEE Transactions on Vehicular Technology* 66 (4): 3293–3305.

37 Xenakis, D., Kountouris, M., Merakos, L. et al. (2016). Performance analysis of network-assisted D2D discovery in random spatial networks. *IEEE Transactions on Wireless Communications* 15 (8): 5695–5707.

38 Hayat, O., Ngah, R., Hashim, Z.Z.M. et al. (2019). Device discovery in D2D communication: a survey. *IEEE Access* 7: 131114–131134.

39 Zhang, J., Deng, L., Li, X. et al. (2017). Novel device-to-device discovery scheme based of random backoff in LTE-advanced networks. *IEEE Transactions on Vehicular Technology* 66 (12): 11404–11408.

40 Li, B., Guo, W., Liang, Y.-C. et al. (2018). Asynchronous device detection for cognitive device-to-device communications. *IEEE Transactions on Wireless Communications* 17: 2443–2456.

41 Kaleem, Z., Khan, A., Hassan, S.A. et al. (2019). Full-duplex enabled time-efficient device discovery for public safety communications. *Mobile Networking Applications* 25: 341–349.

42 Ouali, K., Kassar, M., Nguyen, T.M.T. et al. (2021). Performance evaluation of an effective mobility model for D2D communications. *Wireless Personal Communications* 116: 2675–2696.

43 Lai, W.K., Shieh, C.-S., Chou, F.-S. et al. (2020). Handover management for D2D communication in 5G networks. *Applied Sciences* 10 (12).

44 Kazmi, S.M.A., Khan, L.U., Tran, N.H., and Hong, C.S. (2019). *Network Slicing for 5G and Beyond Networks*. Springer International Publishing.

45 Sun, G., Boateng, G.O., Ayepah-Mensah, D. et al. (2020). Autonomous resource slicing for virtualized vehicular networks with D2D communications based on deep reinforcement learning. *IEEE Systems Journal* 14 (4): 4694–4705.

46 Waqas, M., Niu, Y., Li, Y. et al. (2020). A comprehensive survey on mobility-aware D2D communications: principles, practice and challenges. *IEEE Communication Surveys and Tutorials* 22 (3): 1863–1886.

47 Feki, S., Belghith, A., and Zarai, F. (2020). Multiobjective optimization-based radio resource allocation and sharing algorithm for D2D-based V2V communication. *Transactions on Emerging Telecommunications Technologis* 31: e3783.

48 Hussein, H.H., Radwan, M.H., Elsayed, H.A., and Abd El-Kader, S.M. (2021). Multi V2X channels resource allocation algorithms for D2D 5G network performance enhancement. *Vehicular Communications* 31: 1–12.

49 Ali, K., Nguyen, H.X., Vien, Q.-T. et al. (2021). Review and implementation of resilient public safety networks: 5G, IoT, and emerging technologies. *IEEE Network* 35 (2): 18–25.

50 Ali, K., Nguyen, H.X., Vien, Q. et al. (2020). Deployment of drone-based small cells for public safety communication system. *IEEE Systems Journal* 14 (2): 2882–2891.

51 Rong, H., Wang, Z., Jiang, H. et al. (2019). Energy-aware clustering and routing in infrastructure failure areas with D2D communication. *IEEE Internet of Things Journal* 6 (5): 8645–8657.

52 Masaracchia, A., Nguyen, L.D., Duong, T.Q., and Nguyen, M. (2019). An energy-efficient clustering and routing framework for disaster relief network. *IEEE Access* 7: 56520–56532.

53 Ur Rehman, M., Malik, N.A., Yang, X., and Abbasi, Q.H. (2017). A low profile antenna for millimetre-wave body-centric applications. *IEEE Transactions on Antennas and Propagation* 65 (12): 6329–6337.

54 Andrews, J.G., Bai, T., Kulkarni, M.N. et al. (2017). Modeling and analyzing millimeter wave cellular systems. *IEEE Transactions on Communications* 65 (1): 403–430.

55 Turgut, E. and Gursoy, M.C. (2019). Uplink performance analysis in D2D-enabled millimeter-wave cellular networks with clustered users. *IEEE Transactions on Wireless Communications* 18 (2): 1085–1100.

56 Li, J., Li, X., Wang, A., and Ye, N. (2020). Beamspace MIMO-NOMA for millimeter-wave broadcasting via full-duplex D2D communications. *IEEE Transactions on Broadcasting* 66 (2): 545–554.

57 Soorki, M.N., Saad, W., Manshaei, M.H., and Saidi, H. (2019). Social community-aware content placement in wireless device-to-device communication networks. *IEEE Transactions on Mobile Computing* 18 (8): 1938–1950.

58 Liu, X., Yang, B., Jiang, X. et al. (2020). On social-aware data uploading study of D2D-enabled cellular networks. *Computer Networks* 166.

59 Sun, W., Liu, J., Yue, Y., and Jiang, Y. (2020). Social-aware incentive mechanisms for D2D resource sharing in IIoT. *IEEE Transactions on Industrial Informatics* 16 (8): 5517–5526.

60 Chen, Y., Li, R., Zhao, Z., and Zhang, H. (2021). On the capacity of fractal D2D social networks with hierarchical communications. *IEEE Transactions on Mobile Computing* 20 (6): 2254–2268.

61 Gismalla, M.S.M., Abdullah, M.F.L., Das, B. et al. (2019). Design of an optical attocells configuration for an indoor visible light communications system. *AEÜ - International Journal of Electronics and Communications* 112: 152946.

62 Zhang, H., Ding, W., Yang, F. et al. (2019). Resource allocation in heterogeneous network with visible light communication and D2D: a hierarchical game approach. *IEEE Transactions on Communications* 67 (11): 7616–7628.

63 Gismalla, M.S.M., Abdullah, M.F.L., Niass, M.I. et al. (2020). Improve uniformity for an indoor visible light communication system. *International Journal of Communication Systems* 33 (8): e4349.

64 Becvar, Z., Najla, M., and Mach, P. (2018). Selection between radio frequency and visible light communication bands for D2D, 2018 IEEE 87th Vehicular Technology Conference (VTC Spring), 1–7.

Index

Interference Mitigation in Device-to-Device Communications, First Edition.
Edited by Masood Ur Rehman, Ghazanfar Ali Safdar, and Mohammad Asad Rehman Chaudhry.

www.ingramcontent.com/pod-product-compliance
Lightning Source LLC
LaVergne TN
LVHW062209200926
848484LV00006B/1360
* 9 7 8 1 1 1 9 7 8 8 7 9 9 *